THE HISTORY OF
ENGINEERING SCIENCE

BIBLIOGRAPHIES OF THE HISTORY OF
SCIENCE AND TECHNOLOGY
(VOL. 16)

GARLAND REFERENCE LIBRARY
OF THE HUMANITIES
(VOL. 1150)

Bibliographies of the
History of Science and Technology

Editors

Robert Multhauf, Smithsonian Institution, Washington, D.C.
Ellen Wells, Smithsonian Institution, Washington, D.C.

THE HISTORY OF
ENGINEERING SCIENCE
An Annotated Bibliography

David F. Channell

GARLAND PUBLISHING, INC. • NEW YORK & LONDON
1989

Library of Congress Cataloging-in-Publication Data

Channell, David, 1945–
    The history of engineering science : an annotated bibliography /
David Channell.
        p. cm. — (Bibliographies of the history of science and
technology : vol. 16) (Garland reference library of the humanities ;
vol. 1150)
    Includes index.
    ISBN 0–8240–6636–7 (alk. paper)
    1. Engineering—History—Bibliography. 2. Science—History—
Bibliography. I. Title. II. Series. III. Series: Bibliographies
of the history of science and technology ; v. 15.
Z5851.C47   1989
[TA145]
016.62'0009—dc20   89–34307
                                                                            CIP

Printed on acid-free, 250-year-life paper
Manufactured in the United States of America

In memory of

William Francis Channell

and

Helen Aldous Channell

GENERAL INTRODUCTION

This bibliography is one of a series designed to guide the reader into the history of science and technology. Anyone interested in any of the components of this vast subject area is part of our intended audience, not only the student, but also the scientist interested in the history of his own field (or faced with the necessity of writing an "historical introduction") and the historian, amateur or professional. The latter will not find the bibliographies "exhaustive," although in some fields he may find them the only existing bibliographies. He will in any case not find one of those endless lists in which the important is lumped with the trivial, but rather a "critical" bibliography, largely annotated, and indexed to lead the reader quickly to the most important (or only existing) literature.

Inasmuch as everyone treasures bibliographies it is surprising how few there are in this field. Justly treasured are George Sarton's *Guide to the History of Science* (Waltham, Mass., 1952; 316 pp.), Eugene S. Ferguson's *Bibliography of the History of Technology* (Cambridge, Mass., 1968; 347 pp.), François Russo's *Histoire des Sciences et des Techniques, Bibliographie* (Paris, 2nd ed., 1969; 214 pp.), and Magda Witrow's *ISIS Cumulative Bibliography. A bibliography of the history of science* (London, 1971–; 2131 pp. as of 1976). But all are limited, even the latter, by the virtual impossibility of doing justice to any particular field in a bibliography of limited size and almost unlimited subject matter.

For various reasons, mostly bad, the average scholar prefers adding to the literature, rather than sorting it out. The editors are indebted to the scholars represented in this series for their willingness to expend the time and effort required to pursue the latter objective. Our aim has been to establish a general framework which will give some uniformity to the series, but otherwise to leave the format and contents to the author/compiler. We have urged that introductions be used for essays on "the state of the field," and that selectivity be exercised to limit the length of each volume to the economically practical.

Since the historical literature ranges from very large (e.g., medicine) to very small (chemical technology), some bibliographies will be limited to the most important writings while others will include modest "contributions" and even primary sources. The problem is to give useful guidance into a particular field—or subfield—and its solution is largely left to the author/compiler.

In general, topical volumes (e.g., chemistry) will deal with the subject since about 1700, leaving earlier literature to area or chronological volumes (e.g., medieval science); but here, too, the volumes will vary according to the judgment of the author. The topics are international, with a few exceptions (Greece and Rome, and the United States), but the literature covered depends, of course, on the linguistic equipment of the author and his access to "exotic" literatures.

Robert Multhauf
Ellen Wells
Smithsonian Institution
Washington, D.C.

CONTENTS

PREFACE

The purpose of this volume is to provide students, scholars, and researchers a guide to the history of engineering science. During the past thirty years, historians of science and technology have come to recognize the importance of an intermediate mode of knowledge, labeled engineering science, that lies between science and technology. But most of the existing guides and bibliographies provide only limited information on engineering science. Eugene Ferguson's **Bibliography of the History of Technology** (item 88) lists only seventeen entries under the heading "Engineering Science," while neither the annual bibliography published by the Society for the History of Technology in **Technology and Culture** (item 114) nor the annual bibliography published by the History of Science Society in **Isis** (item 92) contains any separate heading or classification for engineering science. Students and scholars interested in engineering science have also been hindered by the distinctions between the history of science and the history of technology. Works on engineering science that are classified as history of science may go unnoticed by historians of technology and vice versa.

This bibliography has been organized around significant topics in the history of engineering science, and within each section entries are listed alphabetically by author. The first five chapters provide the historical, philosophical, social and institutional framework of engineering science while the last three chapters focus on the specific areas which comprise the traditional core of engineering science. I have specifically excluded works dealing with chemical or electrical technologies. Although many of these works could be classified as contributing to engineering science, they have become more closely associated with the specific disciplines of chemical engineering and electrical engineering. Robert Multhauf has already provided an annotated bibliography on the history of chemical technology, and Bernard Finn is completing an annotated bibliography on the history of electrical engi-

neering as part of the present series. This bibliography also does not cover developments in nuclear engineering, astronautics, laser theory, solid state electronics or computer engineering. Most of these areas
have developed so recently that little historical research has been done on them, and they are changing so
rapidly that any attempt at a bibliographical study at
this time would be doomed to immediate obsolescence.
     Bibliographies such as this are by their nature
selective. This volume does not claim to be a complete
bibliography of the subject, but it does attempt to
serve as a guide to the history of engineering science
and to reflect the wide range of scholarship that is
going on in the field. Both primary and secondary
source materials are contained here. The great majority of the secondary source entries were published after 1958, when the establishment of the Society for the
History of Technology created a growing interest in
professional research in the history of technology. In
most cases those secondary works published before 1958
that are included represent the only significant work
available in that particular subject area. The secondary sources include books, articles and disserta-
tions. Most entries are in English but works in
French, German, Italian, Spanish, Polish, Russian and
Japanese have also been included. The secondary source
entries are the result of an extensive search and survey of the published literature in the history of science and the history of technology as they are most
broadly defined. At times I have made use of other
bibliographies, review essays, book reviews, holdings
of major libraries and footnotes of books and articles
on the subject. A work was included if it made a significant contribution to an understanding of engineering science or if it was representative of the range
and diversity of approaches to the subject. I have
tried to be inclusive rather than exclusive in my judgments.
     Because many historians of science and technology
are unfamiliar with engineering science as a field of
research, I have chosen to include a number of primary
sources which I believe represent the core literature
of engineering science. The majority of these books
and articles were published between 1750 and 1900. I
have tried to note when modern editions are available
and when foreign language works were translated into
English. In order to keep this bibliography within a
manageable size I had to be much more selective concerning primary sources than I had to be with secondary
sources. For example, a complete bibliography of works
by W.J.M. Rankine, an important contributor to the lit-

erature of engineering science, would run to over five
hundred entries, and the same is true for many other
important figures. I selected works that were seminal
ones in the field of engineering science, but also some
lesser known works in order to represent the range of
writing that has contributed to engineering science.
For a more complete list of primary sources, readers
should consult the Royal Society of London's **Catalogue
of Scientific Papers** (see item 107) or Poggendorf's
**Biographisch-literarisches Handworterbuch zür
Geschichte der exakten Wissenschaften** (see item 284),
although many works on engineering are omitted from
these sources.
     Most entries naturally fell into one of the topi-
cal categories but in some cases works could have been
placed under more than one heading. Works that covered
several topics or approaches were usually placed in
Chapter I under the heading "general studies," while
works, especially primary sources, that covered more
than one specific area of engineering science were usually placed in Chapter VI under the heading "general
studies in applied mechanics." At the end of most sections are cross-references to other entries that include a discussion of the same topic. These cross-references refer to citation numbers rather than page numbers. The same is true for the author index at the end
of the book.
     Annotations attempt to describe briefly the con-
tents or main argument of the work, and in many cases
evaluative comments are also included in the annotations. Most of the annotations for primary source materials include some comments on the publication history of the work and on translations. Works without
annotations either were not seen by the author or have
self-explanatory titles.

ACKNOWLEDGMENTS

     In the creation of this bibliography I received a
great deal of both direct and indirect help. I owe a
large debt to Reese Jenkins, Melvin Kranzberg, Edwin
Layton and Robert Schofield who taught me the history
of science and technology. My idea of engineering science has been significantly influenced by the works of
Donald Cardwell, Edwin Layton and Walter Vincenti. I
was also given considerable assistance by the following
institutions and their librarians: the University of
Texas at Dallas, Case Western Reserve University, the
University of Chicago, Southern Methodist University,
the British Library, the National Library of Scotland,
the University of Edinburgh, Glasgow University, and

the University of Cambridge. I must acknowledge the
role of numerous authors and reviewers who, through
their reviews, footnotes and references, alerted me to
books and articles that I might have overlooked. I
must also thank Robert Multhauf for encouraging me in
this project and for being an ideal editor. Finally,
my wife Carolyn provided me with invaluable support
throughout this project.

David F. Channell
The University of Texas at Dallas
Richardson, Texas

INTRODUCTION:
THE HISTORY AND HISTORIOGRAPHY OF ENGINEERING SCIENCE

In the modern world, science and technology have
come to be seen as indistinguishable activities. It is
difficult to classify the work that resulted in the development of the atomic bomb, transistors, lasers, or
the space program as either purely scientific or purely
technological. We assume that scientific discoveries
will lead to some new or improved technology. We also
assume that the successful development of technology
requires a knowledge of science and mathematics. Like
scientists, engineers are trained in universities, tak-
ing many of the same courses in physics, classical mechanics, thermodynamics, calculus and differential
equations that are offered to students majoring in science. Many engineers have degrees in physics or mathematics while many physicists or mathematicians have degrees in engineering. But this interaction between
science and technology that has come to characterize
our modern industrial world is of relatively recent
origin, dating back less than two hundred years.

With some exceptions, science and technology have
followed different social and intellectual traditions,
and have had little impact on each other. Since classical times, science was closely associated with philosophy. As a form of natural philosophy, science was
concerned with asking and answering philosophical questions such as what is the ultimate nature of reality?
With the goal of generating new knowledge about the
world, science put little or no emphasis on using such
knowledge to solve practical problems. Scientists used
mathematical, geometric and other rational methods in
order to idealize problems so that their solutions
could be universally applied to the description of nature as a whole. Scientific knowledge, the result of
contemplation, was also associated with the elite. And
after the establishment of universities during the Middle Ages, most scientists had university degrees. Finally, knowledge of science was gained and disseminated

through a written tradition of treatises, textbooks and journal articles.

Historically, technology has been a practical activity concerned with asking and answering questions about what should be done in particular situations. The goal of technology was to develop workable solutions to practical problems. Engineers traditionally used cut-and-try empiricism and rule-of-thumb techniques that, while applicable to specific situations, could not be generalized to a wide class of problems. Before the nineteenth century, few mechanics or engineers had formalized educations or university training; technology was the province of an artisan class. Knowledge of technology, which resulted from observation and experience, was gained through an apprenticeship system and disseminated through direct contacts between master and pupil. Very little technology was written down.

During the eighteenth and nineteenth centuries, new research and pedagogy arose in such areas as the strength of materials, the theory of machines, the thermodynamics of the steam engine, hydrodynamics and naval architecture. These led to a new scientific approach to technology which could not be easily classified in terms of the traditional categories of science/technology or theory/practice. By the nineteenth century, this new research and pedagogy had become institutionalized in engineering schools, technical institutes and professional societies, and by the second half of the century, it was becoming more and more common to refer to this range of activities under the term engineering science.

THE HISTORY OF ENGINEERING SCIENCE

Although a sustained and institutionalized scientific approach to technology dates back only some two hundred years, there was a long but sporadic history of interactions between science and technology that would pave the way for the emergence of engineering science. The Greeks developed a scientific approach to the construction of catapults while the scientist Archimedes attempted a theory of simple machines based on mechanics. The automatic devices of Hero of Alexandria reflected a knowledge of the science of pneumatics. During the Middle Ages some architects and master masons attempted to systematize the rules of construction, while Hugh of St. Victor argued that mechanics was a mode of knowledge that could be divided into seven

branches, paralleling the seven liberal arts. By the
Renaissance several artist-engineers such as Leonardo
da Vinci provided a means for the interaction of science and technology. Many of these engineers used a
knowledge of science and mathematics to develop theories of machines and structures. Ladislao Reti has
noted that Leonardo's mechanical notebooks show some of
the earliest efforts to develop a theory of machines
based on a study of mechanisms (see items 669 and 670).
     During the Scientific Revolution, new theories of
science led to further contacts with technology. The
emergence of a magical or Hermetic approach to nature
led to a view of science as a tool that could be used
to manipulate the forces of nature to some end or purpose. In his preface to the first English translation
of Euclid, John Dee, a possible model for Shakespeare's
Prospero, indicated that science and such practical activities as architecture, hydraulics, mechanics and
navigation were all governed by principles of geometry,
which he saw as a form of mathematical magic (item
592). At the opposite end of the spectrum from
Hermeticism, mechanical theories of natural philosophy,
which held that nature functioned like a giant machine
or clockwork, encouraged scientists to turn their attention to understanding technology. In his **Discourses
on Two New Sciences** (item 601), Galileo developed a new
dynamics arising out of his investigations of technological problems such as the scale effects in machines
and the strength of materials. To solve problems en-
countered by this emerging experimental tradition
within science scientists were forced to confront technology. For example, they had to understand, use and
design scientific instruments such as the telescope,
microscope, and pendulum clock.
     The recognition of a new relationship between sci-
ence and technology was put forward in the seventeenth
century by Francis Bacon, who argued that the goal of
science was not just knowledge, but knowledge that
would allow humans to exert power and control over nature. Bacon foresaw a technological form of science.
In his **New Atlantis** he described an institution called
Solomon's House in which scientific research was aimed
at improving engines, machines, cannons, clocks, ships,
and other elements of technology. Although Bacon's
goal of a technological science was not realized in his
lifetime, his views influenced the founding of the
Royal Society of London and also set the stage for the
emergence of engineering science during the eighteenth
and nineteenth centuries.
     The idea that a scientific approach to technology
could be institutionalized in a new form of technical

education was offered in the seventeenth century by
Jean-Baptiste Colbert. As chief finance minister to
Louis XIV he believed that French manufacturers and
mercantilists would benefit from educational reform.
At a time when formal education emphasized theology and
the classics, Colbert proposed establishing a number of academies to teach the scientific and mathematical basis of such practical subjects as road and bridge
building, fortification, design, navigation and manu-
facture. Although some of Colbert's plans were put
into place in France during the seventeenth century,
more emphasis was placed on rule-of-thumb techniques
than on mathematical and scientific methods. But his
ideal continued and came to fruition in the following
centuries.

Throughout the eighteenth century most technology
was still based on traditional skills developed through
experience and passed on by means of the apprenticeship
system. But by the end of the century, the ideas of
Bacon and Colbert were beginning to manifest themselves, particularly in France, in the development and
institutionalization of a scientific approach to technological problems. In 1729 Bernard Forest de Bélidor,
a professor at the artillery school at La Feré, published a book entitled **La science de ingénieures** (item
905). The title is notable as one of the first uses of
the term engineering science. In engineering schools
such as the École des Ponts et Chaussées, the École des
Mines, and the École du Génie at Mézières, some of the
first elements of what would be called engineering science emerged. At these schools theoreticians such as
Gaspard Monge, Charles Bossut, Jean Perronent, Baron de
Prony and L.H. Duhamel de Monceau began to develop the
mathematical and scientific principles in such areas as
strength of materials, hydrodynamics, the statics of
structures, and the mechanics of machinery. Some of
the students of these schools such as Lazare Carnot,
Jean Charles Borda, J.V. Poncelet and Charles A. de
Coulomb made significant contributions to engineering
science.

An event of great importance in the history of engineering science was the establishment in Paris of the
École Polytechnique in 1794-95. The school, designed
to train both military and civil engineers, represented
a new recognition that these two branches of technology
depended upon the same principles. At the École Polytechnique, all areas of technology were based on a core
curriculum of geometry, trigonometry, physics, chemistry, mechanics, laboratory and shopwork. Some of the
most famous mathematicians, scientists, and engineers,

including J.L.Lagrange, Simon Laplace, A.L. Cauchy, G.G.
Coriolis, J.M.C., Duhamel, G. Lamé, G. Monge, C.L.M.H.
Navier, S. Poisson, J.V. Poncelet, and G. de Prony,
taught at the school. Through their work, a theoretical approach to technological problems emerged. It was
an area of study based on mathematical techniques and
scientific methodology which could be formalized in
written treatises and articles. The École Polytechnique, by providing a new model for engineering education, helped to spread a scientific approach to technology to other countries. The school's organization
and curriculum influenced Austrian polytechnics at
Vienna and Prague, and German polytechnic schools at
Karlsruhe, Munich, Dresden, Stuttgart, and Hanover. In
the United States, West Point and Rensselaer Polytechnic Institute, two of the earliest engineering schools,
introduced methods and textbooks that had originated at
the École Polytechnique. Many of the leading engineering scientists of the nineteenth century, including
Ferdinand Redtenbacher, Otto Mohr, Franz Grashof,
Friedrich Clebsch, August Föppl, and Ludwig Prandtl,
were associated with these polytechnics.

During the eighteenth and nineteenth centuries,
some engineers came to see the systematic approach provided by science as a way to transform technology from
a craft into a profession. The status of engineers
could be raised not only through the polytechnics but
also through the establishment of other professional
organizations. For example, in Great Britain, elements
of engineering science first emerged from learned societies, self-help groups, and professional institutions.
Formal and informal groups such as the Lunar Society of
Birmingham or the Manchester Literary and Philosophical
Society enabled engineers and industrialists to form
social contacts with scientists, physicians and other
learned individuals. By the beginning of the nineteenth century engineers and mechanics began to form
self-help groups, reading societies, and lending libraries in order to increase their social standing
through education. Most of these mechanics' institutes, as they came to be called, focused on science,
mathematics and elementary engineering science. The
birth of these institutes led, in turn, to publications
such as **The Mechanics' Magazine** and the **Glasgow Mechanics Magazine**, which presented scientific subjects of
interest to engineers and which encouraged a theoretical approach to technology as a way of raising the status of mechanics. An editorial in **The Practical Mechanic and Engineer's Magazine** (1843) argued that the
mechanic who is familiar with theory "becomes, in con-

sequence, not only a more valuable workman, but a more
dignified being than before."[1] By the middle of the
nineteenth century, not only had over six hundred mechanics' institutes been established in Great Britain,
but the movement had spread to America, where the new
Franklin Institute in Philadelphia became a major cen-
ter for research into the area of engineering science.
And in Great Britain, many of the original institutes
eventually underwent re-establishment as technical colleges.

     The issue of the status and role of a scientific
approach to technology was also a factor in the establishment of professional engineering societies. In
Great Britain one of the earliest engineering societies
began in 1771 when a group of engineers and friends of
John Smeaton formed a Society of Civil Engineers, the
forerunner of the Institution of Civil Engineers.
Smeaton had conducted systematic experiments as the basis for his engineering works and was one of the first
to use the title civil engineer. Many of his followers
saw the use of science or theory as an element essen-
tial to the identity of professional engineers and a
means of distinguishing themselves from artisans or mechanics. In England the Institution of Civil Engineers, the Institution of Mechanical Engineers, and the
Institution of Naval Architects encouraged the development of a scientific approach to technology through
meetings, membership standards and publications of
their proceedings and transactions. In the first issue
of **The Engineer**, a journal aimed at the professionalization of engineering in Great Britain, the editor argued that its purpose was "to engage more and more the
practical intellect of the country in the cultivation
of the industrial sciences."[2] In Germany the Verein
Deutscher Ingenieure (the Association of German Engineers), which arose in 1856 from graduates of the
Berlin Institute of Trades, had as one of its main
goals enabling its members to gain social advancement
through scientific training. In America, engineering
societies such as the American Society of Mechanical
Engineers debated whether the professional direction
of engineering would be governed by a "shop culture" or
a "school culture" (see item 533).

     Although a scientific approach to technology
emerged during the late eighteenth century, it did not
begin to have a significant impact on engineering practice until the second half of the nineteenth century.
The advances in technology that contributed to the Industrial Revolution were, for the most part, based on
traditional skills and craftsmanship rather than engi-

neering science. But the new technologies and the increased scale of traditional technologies that resulted
from the Industrial Revolution helped to establish engineering science as a necessary part of technological
development and in the process helped to bring about
its autonomy from pure science. With the development
of the steam engine, railroads, ocean-going iron-hulled
steam ships, large scale suspension bridges and tubular
iron railroad bridges, it became impractical and uneco-
nomical for engineers to rely on traditional rule-of-
thumb or trial-and-error. At the same time, some scientists, who were interested in the practical applications of science, were beginning to discover what most
engineers had always known -- that many of the laws of
science were not directly applicable to technology.
Newtonian mechanics might explain the forces acting between two atoms but did not help to determine how an
iron beam might act under a complex load. Boyle's law,
which explained the relationship between pressure and
volume in an ideal gas, was of little use in describing
how steam acted in a working steam engine. The
Bernoulli equation or the Navier-Stokes' equations of
classical hydrodynamics had limited application in describing real fluids undergoing non-laminar flow.

By mid-nineteenth century engineers were beginning
to recognize engineering science as an autonomous and
systematic activity, whose methods achieved a synthesis
between scientific theory and engineering practice. In
its first editorial, **The Engineer** stated: "There is a
science of the application of science, and one of no
minor importance. The principles of physics ... would
remain only beautiful theories for closet exercise, but
for the science of application."[3] A few months earlier
W. J. Macquorn Rankine, Regius Professor of Civil Engineering and Mechanics at the University of Glasgow, argued that knowledge arising from the harmony of theory
and practice "qualifies the student to plan a structure
or machine for a given purpose, without the necessity
of copying some existing example, and to adapt his designs to situations to which no existing example af-
fords a parallel. It enables him to compute the theoretical limit of the strength or stability of a structure, or the efficiency of a machine of a particular
kind, -- to ascertain how far an actual structure or
machine fails to attain that limit, -- to discover the
causes of such shortcomings, -- and to devise improvements for obviating such causes; and it enables him to
judge how far an established practical rule is founded
on reason, how far on mere custom, and how far on error."[4]

During the second half of the nineteenth century
engineering science began to make significant advances.
Between 1856 and 1861 Max Becker edited a five volume
Handbuch der Ingenieur-Wissenschaft (item 691). In
1862 the University of Glasgow became one of the first
institutions to specifically recognize the autonomy of
engineering science by awarding a Certificate of Profi-
ciency in Engineering Science to students who completed
study in the stability of structures, the strength of
materials, the theory of machines, the thermodynamics
of the steam engine, the principles of hydraulics, and
the mathematics of surveying. Throughout the next
decades engineering science continued to advance and to
develop its own concepts such as stress, strain, efficiency, the Carnot cycle, inverse frames, stream-lines,
boundary-layers, and many others that distinguished engineering science from being simply applied science.
Rather than referring to elements of the natural world,
such as Newtonian forces and atoms, these concepts re-
ferred to idealized elements of a technological world,
such as beams, engines, machines, hulls, and wings.

At the same time some of the first engineering research laboratories were established. In Europe,
Johann Bauschinger created a materials-testing laboratory in 1868 at the Polytechnical Institute of Munich
while in America Robert Thurston founded laboratories
at Stevens Institute of Technology and Cornell University. By the 1870s engineering science also came to
encompass some of the new developments in electricity
and magnetism that were beginning to have an impact on
technology. But as Ronald Kline has shown, electrical
engineering theory differed in several important ways
from the traditional areas of engineering science (see
item 180), and it eventually came to be more closely
associated with the newly emerging discipline of electrical engineering than with engineering science.

In the twentieth century, engineering science has
become an integral part of modern technology. The apprenticeship system has been replaced by university
training in both basic science and engineering science.
Several universities have recognized the importance of
engineering science by creating chaired professorships
and granting degrees in engineering science. In 1919
the Royal Swedish Academy of Engineering Sciences was
founded and in 1963 the Society of Engineering Science
was established in this country and began publishing a
journal entitled **Engineering Science Perspectives.** Recently the basic elements of the field have been codi-
fied in a two-volume **Handbook of the Engineering Sciences** (see item 56). The core of engineering science

continues to be in the areas of applied mechanics,
thermodynamics, and fluid mechanics, but in the past
few years elements of nuclear engineering, astronautics, control systems, solid state electronics, laser
theory, and computer engineering have continued to
broaden and enrich the field.

THE HISTORIOGRAPHY OF ENGINEERING SCIENCE

The past thirty years have seen a growing interest
in the history of engineering science. Before 1960,
only a limited number of works dealt with elements of
engineering science, and their authors were individuals
with training in engineering rather than history. They
adopted what has been called an internalist approach,
which focuses on the emergence, development, and interrelationships of scientific-technological concepts,
theories, and devices. Such internalist studies tend
to minimize the role of social, economic, political,
and cultural forces on the development of science and
technology. But many of these early studies such as
Isaac Todhunter's **A History of the Theory of Elasticity
and the Strength of Materials** (item 896), Henry
Dickinson's **A Short History of the Steam Engine** (item
1190), Stephen Timoshenko's **History of Strength of Materials** (item 897), and Hunter Rouse's and Simon Ince's
History of Hydraulics (item 1360) provided a chronology
and narrative description of important developments in
engineering science; in many cases, they are still the
most significant resources for the history of these elements of engineering science. Also written throughout
this period were a large number of histories of universities, institutes of technology, engineering societies, and schools of engineering. Most of these histories were produced to mark some anniversary; they
were generally narrative, celebratory, and anecdotal in
nature. As a rule, these studies neglect discussing
the internalist development of engineering science concepts, theories and methodologies, but again, many of
them provide the only details of the institutionalization of engineering science that are available in published form. Also, a number of biographical studies of
individuals who contributed to engineering science were
published throughout the nineteenth and twentieth centuries. Many of these were written as obituaries or
memorials and in most cases they are more descriptive
than analytical but in the case of many individuals, no
other biographies exist.

Although these early works dealt with some aspects
of the history of engineering science, almost none of

them used the term engineering science or saw that the
subject matter of their works was an area that was distinguished from either basic science or traditional
technology. In his book **Engineering and Western Civilization** (item 22), James Kip Finch became one of the
first historians to specifically refer to the rise of
engineering science and to see it as a "wedding" of
science and engineering. Finch paved the way for historians to treat works in machine design, engineering
thermodynamics, hydraulics, the mechanics of materials,
and the analysis of structures, as composing the typi-
cal subjects of engineering science and to see that
these subjects shared a common methodological approach
which combined the basic principles of mechanics with
empirical data derived from testing and experimentation.
     After 1960 interest in the history of engineering
science began to grow significantly. A major reason
for this growth was the development of the history of
technology as a professional and scholarly discipline.
In 1958 a group of historians formed the Society for
the History of Technology (SHOT) after meeting with resistance from the History of Science Society when they
proposed including the history of technology under its
purview and publishing articles on technology in its
journal (see item 69). With the establishment of SHOT
and its journal **Technology and Culture**, historians of
technology began to focus their attention on the ways
in which technology was distinct from science. The
second volume of **Technology and Culture** contained a
special issue on the relationship between science and
technology. Several historians of technology began to
challenge the widespread belief, especially among historians of science, that technology was simply applied
science. According to this view, technology was a sub-
discipline of science that did nothing more than apply
the results and discoveries generated by pure science,
without making any fundamental changes or contributions
to those scientific discoveries. Now, however a number
of historians of technology, such as James Kip Finch,
Peter Drucker, Melvin Kranzberg and Cyril Stanley
Smith, supported the position that technology had its
own concepts and methodologies which were independent
of science. Technological knowledge, they argued,
could not be simply reduced to scientific knowledge.
Other historians such as Robert Schofield, Monte
Calvert and Frederick Artz conducted pioneering investigations of the sociological distinctions between science and technology in such institutions as the Lunar
Society of Birmingham (e.g., see item 570), the Ameri-

can Society of Mechanical Engineers (e.g., see item
533), and the French educational system (e.g. see item
419). The issue of the relationship between science
and technology was further fueled by Project Hindsight
(see item 250), undertaken by the Department of De-
fense, which concluded that only a small fraction of
the key contributions to the nation's weapon systems
could be classified as basic or applied science while
the vast majority were classified as technological. In
response the National Science Foundation sponsored a
study entitled TRACES (item 170), which concluded that
five recent innovations had in fact depended on earlier
scientific research. The resulting debate over the relationship between science and technology provided a
new framework for the study of engineering science.
Historians such as Cyril Stanley Smith, Derek Price,
Donald Cardwell, A.R. Hall, Lynwood Bryant and Robert
Multhauf began to distinguish technology from science
in terms of the motivation of doing versus knowing, or
in terms of the production of artifacts versus the production of idealized theories, or in terms of the goal
of creating a structure or machine versus the creation
of a published paper.
     But as many historians came to see important dif-
ferences between science and technology, they also came
to recognize the role of engineering science as an in-
termediary body of knowledge which connects science and
technology. During the 1970s engineering science became an important element of a new "interactive" model
of science and technology. Rather than viewing technology as subordinate to science, the interactive model
assumes a symbiotic relationship in which knowledge,
discoveries, and techniques can flow in both direc-
tions. One of the most influential versions of this
new interactive model was Edwin T. Layton's 1971 paper
"Mirror-Image Twins: The Communities of Science and
Technology in 19th-Century America" (item 202). In
this paper, Layton argued that science and technology
have developed into separate and distinct cultures
which share many of the same values but rank them in
reverse orders. The model of mirror-image twins provided new insights into the history of engineering science. According to Layton, as the theories of elasticity, hydraulics and thermodynamics gradually diverged
from physics, they assumed the mirror-image characteristics of engineering science and provided the technological community with the equivalents of theoretical
and experimental areas of the basic sciences.
     During the first half of the 1970s the framework
provided by the interactive model of science and technology led to a number of significant studies that have

contributed to the history of engineering science.
Some of those studies used the model to shed new light
on the internal history of engineering science. For
example, Donald Cardwell published a prize-winning
study of the rise of thermodynamics in the early industrial age (see item 1174), but unlike previous authors
who focused on either the purely scientific aspects of
thermodynamics or the purely practical problem of the
workings of the steam engine, Cardwell argued that
thermodynamics was a science that emerged from the development of power technologies and owed little or
nothing to previous developments in Newtonian science.

The interactive model also provided a new framework for biographical studies, such as C. Stewart
Gillmor's work on Charles Coulomb (see item 335), and
Charles C. Gillispie's work on Lazare Carnot (see item
333). Previous histories had recognized Coulomb's and
Carnot's contributions to science but often neglected,
or treated as tangential, their interest in technology.
But Gillmor was able to show that Coulomb's intellectual style and approach to problems, including those in
electricity and magnetism, were derived more from engineering science than from pure science. Gillispie
showed that Carnot's contributions to the formulation
of a concept of energy cannot be disentangled from his
study of machines.

The interactive model also provided a useful perspective from which to analyze institutions. For example, Bruce Sinclair's award-winning study of the
Franklin Institute (see item 574) describes the emergence of an experimental research program in such areas
as steam boiler explosions that cannot be analyzed in
terms of either pure science or empirical technology
but must be seen as a science of technology.

Scholars interested in the history of technology
during the medieval and Renaissance periods used the
interactive model to investigate the early roots or
pre-history of engineering science. For example,
Alexander Keller (see items 625-634) showed that sixteenth and seventeenth century engineers such as
Besson, Bronca, Ramelli and Zonca were attempting to
develop a theoretical and systematic study of machines,
while Ladislao Reti (see items 666, 669, and 670) concluded that Leonardo da Vinci's study of machines and
mechanisms in the Madrid Codices was a precursor of
modern engineering theory.

In 1976 Hugh G.J. Aitken provided a fruitful new
way in which to understand the role played by engineering science in the interactive model (see item 118).
He argued that science and technology form distinct

systems that each produce information in a particular
coded form that is not usable in the other system un-
less it is suitably translated into the code of that
system. In this interpretation, engineering science
functions as a translator between pure science and
technology. Although Aitken's notion of translation
was derived from the interactive model, his work also
reflected a growing interest in situating engineering
science in a broader context. He added a new degree of
complexity to the standard interactive model. Aitken
argued that information also flowed back-and-forth, in
translated form, between technology and the economy,
and that "feedback loops" would allow the economic system to affect the translation of information between
science and technology.

The attempt to analyze the interaction of science
and technology from a broader perspective was also reflected in a volume of papers by an international group
of scholars entitled **The Dynamics of Science and Technology: Social Values, Technical Norms and Scientific
Criteria in the Development of Knowledge**, edited by
Wolfgang Krohn, Edwin Layton and Peter Weingart (see
item 38). This set of papers sought to encourage the
development of a new program in which an interdisciplinary approach drawn from the history of science and
technology, sociology, economics, economic history, and
the philosophy of science could be brought to bear on
the problem of the interaction of science and technology. The volume was particularly important because it
sought to develop not only an interdisciplinary approach to the subject but an international one as well.

Also, during the 1970s, another group of historians began to question whether the problem of the interaction of science and technology needed to be reformu-
lated. In 1976 **Technology and Culture** published a special issue based on the proceedings of a Burndy Library
Conference on "The Interaction of Science and Technology in the Industrial Age" (see item 242). Several of
the participants, including Otto Mayr and Arnold
Thackray questioned whether the categories "scientific"
or "technological" were in any sense meaningful. As
John Staudenmaier has noted, this debate led historians
of technology to focus their attention on the nature of
technological knowledge (see item 69).

During the last several years, an increased interest in the nature of technological knowledge has provided a new context for the study of the history of engineering science. One of the most influential studies
of engineering science in terms of technological knowledge was a paper by Edwin Layton presented at the

Burndy Library Conference entitled "American Ideologies
of Science and Engineering" (see item 196). In this
paper, he argued that the statements of engineering
science, unlike those of basic science, were not about
abstract atoms and forces but about idealizations of
machines, beams, heat engines, hulls, and wings.
Through studies with models, wind tunnels, testing machines and towing tanks, engineering science had always
to relate experimental results to actual performances
through attention to scale effects.

    The idea of technological knowledge has served as
an important element in recent studies in the history
of engineering science. In important and insightful
case studies of control-volume analysis, Durand's propeller tests, the Britannia Bridge, flush riveting, and
the design of the Davis wing, Walter G. Vincenti has
investigated how the problem-solving activity associated with technology leads to the knowledge-producing
activity associated with engineering science (see items
262, 263, 1491 and 1492). Like earlier works on the
history of engineering science, Vincenti's studies are
concerned with the relationship between science and
technology, but in his works this concern is sub-
servient to an analysis of the generation of a body of
knowledge and a method of thinking that enables engineers to solve problems.

    By combining the problem of the science-technology
relationship with the problem of the nature of technological knowledge, recent studies have presented sophisticated new approaches for understanding the history of engineering science. One of the most influential approaches has been put forward by Edward W.
Constant, II, in his prize-winning book **The Origins of
the Turbojet Revolution** (see item 1471). In this work
he shows how the values of "scientific" technology and
the culture from which these values were derived led to
the overthrow of traditions of practice and resulted in
the conceptualization of new technological systems.
Drawing on the philosophical studies of Thomas Kuhn,
Karl Popper, Imre Lakatos, and Donald Campbell,
Constant shows how scientific technology could result
in a presumptive anomaly which assumes that although a
conventional system has not actually failed, it would
fail to perform under some future conditions, or that
scientific technology could create new community traditions of testability which would influence technological change. Such concepts as presumptive anomaly and
traditions of testability, along with the idea of technological co-evolution and communities of practition-
ers, have provided historians with new interpretive
tools with which to analyze engineering science.

Constant's work reflects the growing trend in the 1980s to place the study of technological knowledge, and with it the history of engineering science, in a broader philosophical and sociological context. This trend has led to an emphasis on methodological reflection and an increased awareness that models drawn from the philosophy of science, the philosophy of technology, and the sociology of scientific knowledge can be used to inform the study of engineering science. Recently Friedrich Rapp's work on an analytical philosophy of technology has focused on the structure of engineering science as a way to analyze the process of accumulation and self-reinforcement of modern technology (see item 239). A volume of papers entitled **The Nature of Technological Knowledge: Are Models of Scientific Change Relevant?** edited by Rachel Laudan (item 40), has raised new issues concerning the interaction of science and technology by examining the relevance of historical, philosophical and sociological studies of scientific knowledge as exemplified by Thomas Kuhn's **The Structure of Scientific Revolutions**, for a theory of technological change. Most recently, a series of international workshops conducted in Europe has resulted in a set of approaches that could loosely be termed the social construction of technological artifacts or systems (see item 6). This approach has raised new sociological questions concerning the interaction of science and technology in the invention of such things as bakelite or fluorescent lighting; it is concerned with how current theories, tacit knowledge, and design methods that combine to form a "technological frame" are linked to "relevant social groups" so as to bring about a form of closure or stabilization among competing descriptions of a technological artifact or system.

The history of engineering science over the past thirty years has reflected the excitement, rigor, energy, intellectual debate, and wide range of frameworks and methodologies that have arisen from the growing professionalization of the history of science and technology. No single approach should be seen as correct. Rather they all have helped to establish engineering science as an area worthy of serious historical scholarship and they have provided a foundation for generations of future research.

NEEDS AND OPPORTUNITIES

There are still a great many needs and opportunities for future research in the history of engineering science. During the Burndy Library Conference, Arnold Thackray said, "It seems to me that one of the great

tasks that confronts us is to assemble a catalog of ex-
emplary case studies of modes and of levels and types
of interaction between science, however conceived, and
technology, however conceived, at different periods and
in different cultures."[5]   Some work in this area has
been done but Thackray's task still confronts us.   The
history of engineering science remains limited by distinctions and divisions between the history of science
and the history of technology.   The best work will not
be done until historians of technology learn more about
the history of science and historians of science open
themselves to the opportunities that exist in the history of technology.
     The range of areas for new research is great.   The
only monographic studies of the strength of materials
and the history of hydraulics are over thirty years old
and do not make use of recent scholarship.   There are
almost no broad studies on the kinematics of mechanisms
or the theory of machines, and much work needs to be
done on the history of fluid mechanics, including work
in naval architecture and aerodynamics.   Most of the
biographies of major engineering sciences either do not
exist or date back to the nineteenth century, and we
know very little about the individuals who helped create engineering science.   The relationship between engineering science and the rise of engineering education
needs to be explored in much more detail.   A significant number of histories of technical education exist,
but these need to be synthesized with research in the
history of science and technology.   Scholars need to
address the relationship between engineering science
and the professionalization of engineering, and they
need to investigate the role of status in the rise of
engineering science.   Another area where work is just
beginning is cross-cultural studies of engineering science.   How is the rise of engineering science tied to
social, cultural, and philosophical traditions?   How
are national mathematical-scientific traditions such as
the use of infinitesimal calculus in France and the use
of geometrical-graphical techniques in Britain and
America related to the development of engineering science in those areas?   As noted by Edward Constant, II,
a study of traditions of testability in engineering
science should provide fruitful.   Very little research
has been done on the role of instrumentation and experimentation, such as strain gauges, wind tunnels, and
tow tanks, in engineering science.   In spite of the existing significant body of works in the history of engineering science, much opportunity for future research
remains.

Notes

1. **The Practical Mechanic and Engineer's Magazine**, 2 (1843): 2.

2. **The Engineer**, 1 (1856): 3.

3. Ibid.

4. W.J.M. Rankine, **Introductory Lecture on the Harmony of Theory and Practice in Mechanics** (London: Richard Griffin, 1856), pp. 18-19.

5. Nathan Reingold and Arthur Mollela, eds., "The Interaction of Science and Technology in the Industrial Age," **Technology and Culture** 17 (1976): 742.

THE HISTORY OF ENGINEERING SCIENCE: AN ANNOTATED
BIBLIOGRAPHY

CHAPTER I: GENERAL STUDIES AND REFERENCE WORKS

GENERAL STUDIES

1. Armytage, W.H.G. **A Social History of Engineering.** London: Faber and Faber, 1961.

    Presents some useful information on the social and institutional frameworks within which engineering science developed. Includes a brief history of engineering societies.

2. _____. "The Technological Imperative: Scientific Discoveries in the Service of Man." **The 18th Century: Europe in the Age of Enlightenment.** Edited by Albert Cobban. New York: McGraw-Hill, 1969, pp. 95-122.

    Discusses the utilitarian applications of science during the 18th century.

3. Bernal, J.D. **Science and Industry in the 19th Century.** London: Routlege & Kegan Paul, 1953.

    Includes a long essay on the relations between science and technology during the 19th century. Written from a Marxist perspective.

4. _____. **The Social Function of Science.** Cambridge, Mass.: M.I.T. Press, 1967.

    Provides a Marxist interpretation of science. Includes sections on science in education and the application of science to technology. First published in 1939.

5. Bigelow, Jacob. **The Useful Arts, Considered in Connection with the Applications of Science.** New York: Arno, 1972.

Reprint of the 1842 edition. Based on the
Rumford professorship lectures delivered at
Harvard and first published in 1829 as **Elements
of Technology.** One of the earliest works to
treat technology as a body of knowledge. Argues
that technology is dependent on science.

6. Bijker, Wiebe, and Thomas P. Hughes, and Trevor
    J. Pinch, eds. **The Social Construction of
    Technological Systems: New Directions in the
    Sociology and History of Technology.**
    Cambridge, Mass.: M.I.T. Press, 1987.

Provides a new sociological approach to the
study of technology. Includes an examination of
the interaction of science and technology in the
invention of bakelite and flourescent lighting.

7. Birr, Kendall A. "Science in American Industry."
    Science and Society in the United States.
    Edited by David Van Tassel and Michael Hall.
    Homewood, Ill.: The Dorsey Press, 1966, pp.
    35-80.

Discusses the development of technology's de-
pendence upon science during the 19th century.
Focuses on the chemical and electrical industries, but also discusses civil engineering.

8. Borut, Michael. "The **Scientific American** in
    Nineteenth Century America." Ph.D. disserta-
    tion. New York University, 1977.

Analyzes the role of the **Scientific American**
in 19th century America. Concludes that the
journal was influential in disseminating scientific ideas to mechanics through its ideology of
self-help.

9. Braun, Hans-Joachim. "Methodenprobleme der
    Ingenieurwissenschaft, 1850 bis 1900."
    Technikgeschichte 44 (1977): 1-18.

Discusses the problem of methodology in engineering science during the last half of the 19th
century.

10. Bruce, A.K. "State of the Art and Science in
     1856." **The Engineer Centenary Number** (1956):
     154-59.

Useful survey of the role of science in engineering during the mid-19th century.

11.   Bugliarello, George, and Dean B. Donner, eds.
      The History and Philosophy of Technology.
      Urbana: University of Illinois Press, 1979.

Papers by Arthur Donovan and Harold Burstyn
discuss the relationship between science and
technology in the invention of Watt's separate
condenser. Paper by Donald Cardwell argues that
the distinctions between force, energy and power
were the result of developments in 19th century
technology.

12.   Burnett, John Nicholas, ed. **Technology and Sci-
      ence: Important Distinctions for the Liberal
      Arts Colleges.** Davidson, N.C.: Davidson-
      Sloan New Liberal Arts Program, 1984.

Proceedings of a conference sponsored by the
Sloan Foundation and Davidson College on the New
Liberal Arts Program. Contains several essays on
the relationship between science and technology.
Includes items 123, 135, 198, 1223.

13.   Burstall, Aubrey F. **A History of Mechanical En-
      gineering.** London: Faber and Faber, 1963.

Surveys mechanical engineering from 3000 B.C.
to 1960 A.D. In the later period he discusses
the role of science on the development of mechanical engineering. Rather sketchy coverage.

14.   Calhoun, Daniel H. **Professional Lives in Amer-
      ica.** Cambridge, Mass.: Harvard University
      Press, 1965.

Includes an analysis of the professionalization of civil engineers in America.

15.   Cardwell, Donald S.L. **Turning Points in Western
      Technology: A Study of Technology, Science
      and History.** New York: Science History Pub-
      lications, 1972.

Argues for a history of technology related to
the history of science and the history of ideas.
Believes that technology is an equal partner with
science. Covers the developments from medieval
and Renaissance inventions of the mechanical

clock, printing and navigation to the science-
based technology derived from Bacon and Galileo,
to the Industrial Revolution, and to the modern
industrial research laboratory. Discusses the
history of mechanics, strength of materials,
thermodynamics, hydrodynamics, and electromag-
netic field theory. Originally published in England as **Technology, Science and History: A Short
Study of the Major Developments in the History of
Western Mechanical Technology and Their Relationships with Science and Other Forms of Knowledge.**
London: Heinemann Educational, 1972.

16.   Crosland, M. "Science and the Franco-Prussian
       War." **Social Studies of Science** 6 (1976):
       185-214.

       Discusses the application of science to the
area of military engineering during the Franco-
Prussian war.

17.   Daniels, George H. **American Science in the Age
       of Jackson.** New York: Columbia University
       Press, 1968.

       Discusses the relationship between American
science and Baconian ideals about the practicality of science. Traces such ideas to Scottish
Common Sense philosophy and Protestant theology.
Does not directly analyze the relationship between science and technology.

18.   Daumas, Maurice, ed. **Histoire générale des
       techniques.** 4 vols. Paris: Presses universi-
       taires de France, 1962-.

       Survey of Western technology. Includes dis-
cussions of the influence of science on technology. First three volumes published to date.
Also available in English translation.

19.   **Engineering Heritage.** London: Heinemann for the
       Institution of Mechanical Engineers, 1963.

       Contains articles on water driven prime
movers, hydraulic power transmission, oil engines, and the steam engine.

20.   Ferguson, Eugene S. "Elegant Inventions: The
       Artistic Component of Technology." **Technol-
       ogy and Culture** 19 (1978): 450-460.

Argues that the artistic concept of design is as important, or more important, to the development of technology than is science. Focuses on the rocking beam of a Newcomen engine and the construction of wooden toys from a ring as examples of elegant inventions.

21.   _____. "On the Origin and Development of American Mechanical 'Know-How.'" **Midcontinent American Studies Journal** 3 (1962): 3-15.

Argues that the development and spread of technical knowledge through literature, actual machines and skilled craftsmen were an important factor in American technological development.

22. Finch, James Kip. **The Story of Engineering**. Garden City, N.Y.: Anchor Books, 1960.

Surveys the evolution of engineering from a craft to a developed science. Focuses on bridges, machines, canals, roads and dams. One of the first books to focus on engineering science as a specific field. Includes a chapter on "The New Engineering: The Rise of Engineering Science."

23. FitzSimons, Neal, ed. **Engineering Classics of James Kip Finch**. Kensington, Md.: Cedar Press, 1978.

Contains a collection of thirty-four articles on "Great Books of Engineering" by James Kip Finch, originally published in the **Consulting Engineer**. Focuses on civil engineering from the Rhind Papyrus of ancient Egypt through works in the 19th century. Includes works on Agricola, Ramelli, Gautier, Belidor, Smeaton, Rennie, Tredgold and Rankine.

24. Gille, Bertrand, et al. **Histoire des Techniques**. 2 vols. Paris: Gallimard, 1978.

Useful survey of the history of technology. Contains a section on science and technology. The second volume focuses on techniques and science. Translated into English as **History of Techniques**. New York: Gordon & Breach, 1986.

25. Gillispie, Charles C., ed. **A Diderot Pictorial Encyclopedia of Trades and Industry: Manufacturing and the Technical Arts in Plates Selected from "L' Encylopedie, ou Dictionnaire Raissonné des Sciences, des Arts et des Métiers" of Denis Diderot**. 2 vols. New York: Dover Publications, 1959.

Shows how Diderot used science as an instrument of power. Argues that Diderot was influenced by Francis Bacon rather than Newton. Gives examples of how science was applied to technology.

26. _____. **Science and Polity in France at the End of the old Regime**. Princeton: Princeton University Press, 1980.

Argues that science played a different role in the industrialization of France than it did in England. Concludes that science in France was more centralized and more professionalized resulting in fewer direct contacts between scientists and technologists. Rather the government and the military caused science's spread to industry.

27. Hindle, Brooke. **Emulation and Invention**. New York: New York University Press, 1981.

Argues that two important American inventions, the steamboat and the telegraph, involved spatial thinking (synthetic and holistic) as compared to verbal thinking (analytic and sequential). Argues that Robert Fulton's and Samuel F.B. Morse's artistic background led to these inventions.

28. ————, and Melvin Kranzberg, eds. **Bridge to the Future: A Centennial Celebration of the Brooklyn Bridge**. New York: New York Academy of Sciences, 1984.

Contains useful essays on the relationship between science and technology and the role of design in engineering.

29. Hughes, Thomas Park, ed. **The Development of Western Technology since 1500**. New York: Macmillan, 1964.

Good survey of the history of technology, including the role of science.

30. _____. "Emerging Themes in the History of Technology." **Technology and Culture** 20 (1979):
697-711.

Provides a very useful and thoughtful
overview of themes that have emerged in the history of technology. Includes a discussion of the
interaction of science and technology as a theme.

31. Hutchings, Raymond. **Soviet Science, Technology,
Design: Interaction and Convergence.**
London: Oxford University Press, 1976.

Contends that Soviet science before 1917 was
more developed than technology, which was mostly
limited to military and state interests. Notes
that technology is more likely to be influenced
by political and ideological factors. Devotes
only one short concluding chapter to the interaction and convergence of science and technology.

32. Jewkes, John, David Sawers, and Richard
Stillerman. **The Sources of Invention**. 2d
rev. ed. New York: Norton, 1969.

Documents the history of invention through
case studies. Concludes that the theory that inventions arise out of advances in pure science
does not provide the full story of modern inventions.

33. Kargon, Robert H. **Science in Victorian
Manchester**. Baltimore: Johns Hopkins University Press, 1978.

Describes the evolution of Manchester science
in response to scientific, technological and social pressures. Describes how a gentlemen's science was taken over by a utilitarian science.
Analyzes the institutionalization of science in
the university.

34. Klemm, Friedrich. **A History of Western Technology**. Translated by D.W. Singer. Cambridge,
Mass.: M.I.T. Press, 1968.

Useful survey of Western technology based on
a collection of historical documents with commen-

tary. Contains a useful section on scientific
engineering. Originally published as **Technik;
eine Geschichte ihrer Probleme**. Freiburg: Alber,
1954.

35. Koessler, Paul. "Bildungswerte der Technik."
VDI-Zeitschrift 110, no. 5 (1968): 161-66.

Argues for an education in the engineering
sciences. Concludes that it promotes the development of logical thinking.

36. Kranzberg, Melvin, and William H. Davenport, eds.
Technology and Culture: An Anthology. New
York: New American Library, 1972.

Useful anthology of articles reprinted from
the journal **Technology and Culture**. Several articles discuss the relationship between science
and technology.

37. _____, and Carroll W. Pursell, Jr., eds. **Technology in Western Civilization**. 2 vols. New
York: Oxford University Press, 1967.

Widely used survey of technology in Western
civilization. Each chapter is written by a spe-
cialist in the field. Engineering science is included in chapters on the rise of modern civil
engineering, the steam engine, energy conversion,
buildings and construction, the internal combustion engine, and industrial research, among
others.

38. Krohn, Wolfgang, and Edwin T. Layton, Jr., and
Peter Weingart, eds. **The Dynamics of Science
and Technology** (Sociology of the Sciences: A
Yearbook, Volume 2). Dordrecht and Boston:
D. Reidel, 1978.

Contains several essays on the interaction of
science and technology. Includes items 119, 125,
172, 176, 193, 201, 268, 478.

39. Landes, David. **The Unbound Prometheus: Techno-
logical Change and Industrial Development in
Western Europe from 1750 to the Present**.
Cambridge: Cambridge University Press, 1969.

Useful survey of European technological developments with emphasis on the role of eco-

nomics. Contains some discussion of the impact
of science on technology.

40. Laudan, Rachel, ed. **The Nature of Technological
     Knowledge: Are Models of Scientific Change
     Relevant?** Dordrecht and Boston: D. Reidel,
     1984.

     Contains a useful set of papers on the rela-
     tionship between science and technology. In-
     cludes items 141, 159, 168, 195, 235, 269.

41. Layton, Edwin T., Jr., ed. **Technology and Social
     Change in America.** New York: Harper & Row,
     1973.

     Collection of edited reprints, without foot-
     notes, of articles. Includes Eugene S. Ferguson
     on "Technology as Knowledge," Bruce Sinclair on
     "The Direction of Technology," Carl W. Condit on
     "Science and Technology," and Edwin T. Layton,
     Jr. on "Engineers in Revolt," among others.
     Aimed at the undergraduate level.

42. Malkin, I. "On the Power of Scientific Tradi-
     tion." **Scripta Mathematica** 26 (1961): 339-
     46.

     Discusses the engineering science tradition
     in Russia.

43. Marks, John. **Science and the Making of the Mod-
     ern World.** London: Heinemann, 1983.

     Includes some discussion of the rise of
     thermodynamics and its relationship to the steam
     engine.

44. Mason, Stephen F. **A History of the Sciences.**
     rev. ed. New York: Collier Books, 1962.

     Useful survey of the history of science.
     Contains chapters on "Scientific Institutions in
     France and Britain during the Nineteenth Cen-
     tury," "Thermodynamics: The Science of Energy
     Changes," and "Science and Engineering." Earlier
     version published as **Main Currents of Scientific
     Thought.**

45. Mayr, Otto, ed. **Philosophers and Machines.** New
     York: Science History Publications, 1976.

Collection of reprinted articles from the
journal **Isis**. Includes Henry J. Webb's "The Science of Gunnery in Elizabethan England," Donald
Fleming's "Latent Heat and the Invention of the
Watt Engine," Milton Kerker's "Sadi Carnot and
the Steam Engine," Thomas Kuhn's "Sadi Carnot and
the Cagnard Engine," Hilda L. Norman's "Leopardi
and the Machine," and Otto Mayr's "Maxwell and
the Origin of Cybernetics," among others.

46.   Merdinger, Charles J. **Civil Engineering through
      the Ages**. New York: Society of American Mil-
      itary Engineers, 1963.

Surveys the history of civil engineering from
ancient times. Defines engineering as the application of science. Comes to some questionable
conclusions concerning the early history of engineering.

47.   Morison, Elting E. **From Know-How to Nowhere: The
      Development of American Technology**. New
      York: Basic Books, 1974.

Traces the history of American technology
from 1800 to the twentieth century. Compares the
early use of a rule-of-thumb approach with the
use of theoretical and scientific knowledge.
Concludes that technological knowledge must be
linked to the proper use of technology.

48.   Mumford, Lewis. **Technics and Civilization**. New
      York: Harcourt, Brace & World, 1934.

Classic survey of the history of technology.
Classifies the development of a scientific approach as the "neotechnic" phase of technological
development.

49.   Musson, Albert E., ed. **Science, Technology, and
      Economic Growth in the 18th Century**. London:
      Methuen, 1972.

A series of essays, all previously published
elsewhere. Most essays discuss the role of science in the Industrial Revolution.

50.   Noble, David F. **America by Design: Science,
      Technology, and the Rise of Corporate Capi-
      talism**. New York: Alfred A. Knopf, 1977.

A Marxist interpretation of the interaction
of science and technology. Argues that engineering schools provided the new managerial elite
which shaped industrial development. Shows that
new science-based industries depended on the engineering schools to provide knowledge and exper-
tise for new products and processes.

51. Pacy, Arnold. **The Culture of Technology.**
    Cambridge, Mass.: M.I.T. Press, 1983.

    Includes some discussion of the interaction
of science and technology.

52. _____. **The Maze of Ingenuity.** Cambridge,
    Mass.: M.I.T. Press, 1974.

    Useful discussion of the process of invention
and creativity. Includes a section on the effects of mechanisms from 1810 to 1870. Compares
European and Asian technology.

53. Parsons, Charles A. "Engineering Science Before,
    During and After the War." **Science** 50
    (1920): 333-38.

    Presidential Address of the British Associa-
tion for the Advancement of Science in 1919 by a
leader in engineering science.

54. Petroski, Henry. **To Engineer Is Human: The Role
    of Failure in Successful Design.** New York:
    St. Martin's Press, 1985.

    Presents a case-by-case study of recent engi-
neering failures in order to investigate the role
of design. Distinguishes engineering from scientific inquiry. Argues that engineering design
functions as a hypothesis which is tested against
a series of failure criteria. Does not discuss
how engineering research advances design.

55. Plum, Werner. **Natural Science and Technology on
    the Road to the Industrial Revolution.** Bonn-
    Bad Godesberg: Friedrich-Ebert-Stiftung,
    1974.

    Discusses the influence of science and tech-
nology on the Industrial Revolution.

56. Potter, James H., ed. **Handbook of the Engineer-
     ing Sciences.** 2 volumes. Princeton:
     Princeton University Press, 1967.

     Technical handbook aimed at engineers. Sur-
     veys the modern concept of engineering science.

57. Price, Derek J. de Solla. **Science since Babylon.**
     New Haven: Yale University Press, 1961.

     Contains five essays on problems in science
     and technology. The essay on "The Renaissance
     Roots of Yankee Ingenuity" argues that American
     science developed out of Yankee mechanics who
     worked in the tradition of European
     "practitioners."

58. Pursell, Carroll W., Jr., ed. **Readings in Tech-
     nology and American Life.** New York: Oxford
     University Press, 1969.

     Provides useful primary source essays on
     American technology. Includes works by Oliver
     Evans on "Steam Power" and Timothy Walker on a
     "Defense of Mechanical Philosophy." Also con-
     tains a section on the institutionalization of
     engineering including mechanics' institutes and
     university education.

59. _____, ed. **Technology in America: A History of
     Individuals and Ideas.** Cambridge, Mass.:
     M.I.T. Press, 1981.

     Includes articles by Darwin Stapleton on
     "Benjamin Henry Latrobe and the Transfer of Tech-
     nology," Bruce Sinclair on "Thomas P. Jones and
     the Evolution of Technical Education," Edwin T.
     Layton, Jr. on "James B. Francis and the Rise of
     Scientific Technology," Reese V. Jenkins on
     "George Eastman and the Coming of Industrial Re-
     search in America," and Barton C. Hacker on
     "Robert H. Goddard and the Origins of Space
     Flight," among others. Aimed at the undergradu-
     ate level.

60. Rae, John B. "The 'Know-How' Tradition: Tech-
     nology in American History." **Technology and
     Culture** 1 (1960): 139-150.

     Argues that the pragmatic character of Ameri-
     can life placed a high value on the direct prac-

tical applications of science. Concludes that
there was little interest in accumulating scientific knowledge for future technological advances
or for its own sake.

61. Rapp, Friedrich, ed. **Contributions to a Philoso-
     phy of Technology: Studies in the Structure
     of Thinking in the Technological Sciences.**
     Dordrecht, Holland: D. Reidel, 1974.

     Contains reprints of articles previously pub-
     lished including some that appeared in **Technology
     and Culture.** Also includes English translations
     of several other important articles published in
     Czech, German and Russian. Focuses on the rela-
     tionship between science and technology. In-
     cludes items 72, 174, 216, 240, 259.

62. Rolt, L.T.C. **Victorian Engineering.** London:
     Allen Lane, 1970.

     Provides an overview of Victorian engineer-
     ing. Discusses the development of steam engines,
     ships, civil engineering, and electrical power.
     Gives useful biographical details but the histor-
     ical framework is unsophisticated.

63. Russell, Colin A., and D.C. Goodman, eds. **Sci-
     ence and the Rise of Technology since 1800.**
     New York: Harper, Row, 1973.

     Prepared by the faculty of the Open Univer-
     sity in England as a second level course.

64. Russo, François. **Introduction à l'histoire des
     techniques.** Paris: Blanchard, 1986.

     Useful introduction to the history of tech-
     nology. Includes some discussion of the rela-
     tionship between science and technology.

65. Schneider, Ivo. "Die mathematischen Praktiker im
     See- Vermessungs- und Wehrwesen vom 15. bis
     zum 19. Jahrhundert." **Technikgeschichte** 37
     no. 3 (1970): 211-242.

     Discusses the application of mathematics to
     navigation, surveying and military engineering.

66. Singer, Charles, and E.J. Holmyard, and A.R.
     Hall, and Trevor I. Williams, eds. **A History
     of Technology.** 5 vols. New York: Oxford Uni-
     versity Press, 1954-58.

     Monumental survey of the history of technol-
     ogy from ancient times to 1900. Useful as an en-
     cyclopaedia of facts but weaker on interpreta-
     tion. Includes chapter containing some factual
     information on the interaction of science and
     technology and on engineering science.

67. Smith, Cyril Stanley. **A Search for Structure:
     Selected Essays on Science, Art, and History.**
     Cambridge, Mass.: M.I.T. Press, 1981.

     Argues that technological problems have stim-
     ulated scientific theories and experiments. Col-
     lection of fourteen essays.

68. Sporn, Philip. **Foundations of Engineering.** New
     York: Macmillan, 1964.

     Includes a chapter on "Philosophy of Engi-
     neering." Argues that the engineer translates
     scientific knowledge into tools, energy and
     labor.

69. Staudenmaier, John M. **Technology's Storytellers:
     Reweaving the Human Fabric.** Cambridge,
     Mass.: M.I.T. Press, 1985.

     Provides an analysis of the patterns of val-
     ues which have existed in twenty-five years of
     published research in the history of technology.
     Includes a chapter on "Science, Technology, and
     the Characteristics of Technological Knowledge."
     Argues that the debate over the interaction of
     science and technology has shifted to a study of
     the nature of technological knowledge.

70. Straub, Hans. **A History of Civil Engineering.**
     Translated by Erwin Rockwell. Cambridge,
     Mass.: M.I.T. Press, 1964.

     Useful survey of civil engineering. Empha-
     sizes the role of science and theory in the de-
     velopment of civil engineering. Originally pub-
     lished as **Die Geschichte der Bauingenieurkunst.**
     Basel: Verlag Birkhauser, 1949.

71. Struik, Dirk J. **Yankee Science in the Making.**
    rev. ed. New York: Collier Books, 1962.

    Marxist interpretation of the development of
    science and technology in New England before the
    Civil War. Argues that science developed out of
    the interests of the mercantile class living
    along the Eastern seaboard while technology de-
    veloped out of the interest of the farmers in the
    interior of the country. Concludes that science
    and technology began to influence each other when
    industrialist interests replaced mercantile in-
    terests.

72. Teichmann, D. "On the Classification of the
    Technological Sciences." **Contributions to a
    Philosophy of Technology: Studies in the
    Structure of Thinking in the Technological
    Sciences** (item 61), pp. 134-39.

    Analyzes how the technological sciences
    should be classified. Puts forward four possi-
    bilities: historical classification, division ac-
    cording to natural laws, division according to
    branches of production, and division according to
    function.

73. Thackray, Arnold, ed. **Contemporary Classics in
    Engineering and Applied Science.** Foreword by
    Melvin Kranzberg. Preface by Eugene
    Garfield. Philadelphia: ISI Press, 1986.

74. Timm, Albrecht. **Kleine Geschichte der
    Technologie.** Stuttgart: W. Kohlhammer
    Verlag, 1964.

    Focuses on Johann Beckmann's 18th-century
    definition of technology as the science of trans-
    forming raw materials into useful products. In-
    cludes a section on the history of the science of
    technology. Emphasizes German technology.

75. Troitzsch, Ulrich, and Gabriele Wohlauf, eds.
    **Technik-Geschichte: Historische Beiträge und
    neuere Ansätze.** Frankfurt am Main: Suhrkamp,
    1980.

    A set of essays focusing on the history of
    technology in Germany. Includes articles on the
    rationality of fortress design and on the reform

of engineering education at the end of the 19th
century.

76. Truesdell, Clifford. **Essays in the History of
     Mechanics.** New York: Springer-Verlag, 1968.

     Includes essays on the mechanics of Leonardo,
     mechanics in the Age of Reason, the evolution of
     the concept of stress, and recent developments in
     rational mechanics. Links the idea of stress to
     work in hydraulics and hydrodynamics.

77. Usher, Abbot Payson. **A History of Mechanical In-
     vention.** Boston: Beacon Press, 1959.

     Surveys mechanical invention from antiquity
     to the twentieth century. Includes a chapter on
     the early history of the pure and applied mechan-
     ical sciences.

78. Weber, Daniel B. **"The Manufacturer and Builder:**
     Science, Technology, and the American Me-
     chanic." **Journal of American Culture** 8
     (April,1985): 35-42.

     Analyzes the 19th-century American journal
     The Manufacturer and Builder.

79. Wolf, Abraham. **A History of Science, Technology,
     and Philosophy in the 16th and 17th Cen-
     turies.** 2d ed. London: Allen & Unwin, 1950.

     Massive survey of science and technology in
     the 16th and 17th centuries. Particularly useful
     in the prehistory of engineering science.

80. _____. **A History of Science, Technology, and
     Philosophy in the Eighteenth Century.** 2d ed.
     London: Allen & Unwin, 1952.

     Massive survey of 18th-century science and
     technology. Includes material on the interaction
     of science and technology.

See also: 605, 606, 620, 649, 651.

REFERENCE WORKS

81. American Society of Mechanical Engineers, Gas
     Turbine Power Division. **Bibliography on Gas
     Turbines, 1896-1948.** New York: American So-
     ciety of Mechanical Engineers, 1962.

     Includes 200 titles on gas turbines.

82. Besterman, Theodore. **Technology, Including
     Patents: A Bibliography of Bibliographies.** 2
     volumes. Totowa, N.J.: Rowman & Littlefield,
     1971.

     Composed of references to technology and sci-
     ence in Besterman's **World Bibliography of Bibli-
     ographies**, 4th edition.

83. Bradley, Margaret. "An Early Science Library and
     the Provision of Textbooks: The Ecole Poly-
     technique, 1794-1815." **Libri** 26 (1976): 165-
     80.

     Discusses the textbooks used at the École
     Polytechnique.

84. Chrimes, Mike. "Bridges: A Bibliography of Arti-
     cles Published in Scientific Periodicals,
     1800-1829." **History of Technology** 10 (1985):
     217-257.

     Useful bibliography of the theory of bridge
     building in the early 19th century.

85. Durbin, Paul T., ed. **A Guide to the Culture of
     Science, Technology, and Medicine.** New York:
     Free Press, 1980.

     Highly idiosyncratic essays and bibliog-
     raphies covering the history of science, history
     of technology, philosophy of science, philosophy
     of technology, sociology of science and technol-
     ogy, and the history, philosophy, and sociology
     of medicine. Essays include a historical
     overview of each area.

86. École des Ponts et Chaussées. **Catalogue des
     livres composant la bibliotheque de l'École
     des ponts et chaussées.** Paris, 1872.

Catalogue of books in the library of one of
the leading engineering schools in France.

87. _____. **Catalogue des manuscripts de la
      bibliotheque de l'École des ponts et
      chaussées.** Paris, 1886.

Catalog of manuscripts in the library of one
of the leading engineering schools in France.
Includes the papers of several important engineering scientists.

88. Ferguson, Eugene S. **Bibliography of the History
      of Technology.** Cambridge, Mass.: M.I.T.
      Press, 1968.

Standard bibliography in the history of tech-
nology against which all others are measured.
Includes both primary and secondary sources.
Contains a chapter on technical societies and ed-
ucation and a section on "Engineering Sciences."
Invaluable tool.

89. Goodstein, Judith R., and Carolyn Kopp, eds. **The
      Theodore von Karman Collection and the Cali-
      fornia Institute of Technology Guide to the
      Original Collection and a Microfilm Edition.**
      Pasadena: Institute Archives, Robert A.
      Milliken Memorial Library, California Insti-
      tute of Technology, 1981.

Provides a guide to the Von Karman collection. Particularly useful for the history of
aerodynamics and fluid dynamics.

90. **Guide to the Smithsonian Archives.** Washington,
      D.C.: Smithsonian Institution Press, 1983.

Provides access to the Smithsonian's archive
collection. Includes institutional records,
manuscripts of scientists and engineers, and private papers of Smithsonian affiliates.

91. Hindle, Brooke. **Technology in Early America:
      Needs and Opportunities for Study.** Chapel
      Hill: University of North Carolina Press,
      1966.

Contains a very useful bibliography of early
American technology, including both primary and

secondary sources. Also includes a chapter on
"Education, Organization, and Science."

92. Isis. "Critical Bibliography of the History of
    Science and Its Cultural Influences." **Isis**,
    1913-.

    Published annually by the History of Science
    Society. Divided according to scientific disci-
    pline and chronology. Works on engineering sci-
    ence can be found under general works, technol-
    ogy, physical sciences, and scientific education
    headings. Includes a name and author index.
    Most recently edited by John Neu. Important
    tool.

93. _____. **Isis Cumulative Bibliography. A
    Bibliography of the History of Science Formed
    from Isis Critical Bibliographies 1-90, 1913-
    1965**. Volume I: **Personalities, A-J**. Volume
    II: **Personalities, K-Z, and Institutions**.
    Volume III: **Subjects**. London: Mansell, 1971-
    76.

    Brings together the individual yearly criti-
    cal bibliographies into a cumulative bibliog-
    raphy. Particularly useful on personalities and
    institutions. A fourth volume will focus on
    chronological periods. Edited by Magda Whitrow.
    Has been continued into the more recent period by
    John Neu, indexing the literature from 1965 to
    1974. Volume I: **Personalities and Institutions**.
    London: Mansell, 1980.

94. Jayawardene, S.A. **Reference Books for the Histo-
    rian of Science: A Handlist**. London: Science
    Museum Library, 1982.

    Contains bibliographies, guides, biogra-
    phies, archives, library catalogues and other in-
    formation. Includes materials on technology as
    well as science.

95. Josephson, A.G.S., comp. **A List of Books on the
    History of Industry and Industrial Arts**.
    Chicago: John Crerar Library, 1915.

    Provides a guide to books in the John Crerar
    Library relating to technology. The Library's
    collections in science and technology are some of
    the best in the world.

96. McCabe, Irena M., and Frank A.J.L. James.
     "History of Science and Technology Resources
     at the Royal Institution of Great Britain."
     British Journal for the History of Science 17
     (1984): 205-209.

     Contains some useful information on the re-
     sources of the Royal Institution.

97. Mauel, Karl. "Technikgeschichte in ingenieur-
     wissenschaftlichen Werken des 19. Jahr-
     hunderts." **Technikgeschichte** 50 (1983): 289-
     305.

     Presents a summary of the historical writings
     in 18th- and 19th-century engineering science.

98. Meixner, Esther Chilstrom. **Guide to the Micro-
     film Edition of the John Ericsson Papers.**
     Philadelphia: American Swedish Historical
     Foundation, 1970.

     Particularly useful for the history of naval
     architecture and steam power.

99. Mitcham, Carl, and Robert Mackey. "Bibliography
     of the Philosophy of Technology." **Technology
     and Culture** 14 (April 1973): part II.

     Useful bibliography of the role of technology
     in society. Many entries are annotated. Section
     on "Metaphysical and Epistemological Studies" in-
     cludes several entries on the relationship be-
     tween science and technology. Appendix includes
     historical and historico-philosophical studies
     which touch upon engineering science.

100. Newcomen Society. "Analytical Bibliography of
     the History of Engineering and Applied Sci-
     ence." **Transactions of the Newcomen Society**,
     1921-.

     Useful bibliography of publications in the
     history of technology. Published yearly by the
     Newcomen Society.

101. Prokter, C.E., comp. **"The Engineer" Index, 1856-
     1959.** London: Morgan Bros., 1964.

Useful index to an important British engineering journal. Includes both name and subject
indices.

102. **Pure and Applied Science Books, 1896-1980.** Pre-
       pared by R.R. Bowker Company's Department of
       Bibliography, in collaboration with the Pub-
       lication Systems Department. 6 vols. New
       York and London: R.R. Bowker, 1982.

Includes over 220,000 entries of works published in the United States. Introduction includes 5000 works published before 1876. Classification scheme is extremely detailed, comprising
56,000 subject headings.

103. Rink Evald. **Technical Americana: A Checklist of
       Technical Publications Printed before 1831.**
       Foreword by Eugene S. Ferguson. Millwood,
       N.Y.: Kraus International Publications, 1981.

Provides a list of over 6,000 technical works
published in American before 1831. Includes lo-
cations of known copies. Also includes a section
on "Sciences Applied to Technology."

104. Rothenberg, Marc. **The History of Science and
       Technology in the United States: A Critical
       and Selected Bibliography.** New York and
       London: Garland Publishing, 1982.

Contains over 800 annotated entries dealing
with science and technology in the United States.
Includes sections on the science-technology relation, and on science, technology and education.

105. Rouse, Hunter. **Historic Writings on Hydraulics:
       A Catalogue of the History of Hydraulics Col-
       lection in the University of Iowa Libraries.**
       Iowa City: Friends of the University of Iowa
       Libraries, 1984.

Useful introduction to the resources in the
history of hydraulics by one of the leading
scholars in the field.

106. Royal Institution of Great Britain. **The Archives
       of the Royal Institution of Great Britain.
       Minutes of Managers' Meetings, 1799-1900.** 2
       volumes. Menston, Ilkley, Yorkshire: Scolar
       Press, 1971.

Presents the role of technology in the Royal
Institution.

107. Royal Society of London. **Catalogue of Scientific
      Papers, 1800-1900.** Vols. 1-7. London: HMSO,
      1867-77; Vol. 8. London: John Murry, 1879;
      Vols. 9-19. Cambridge: Cambridge University
      Press, 1891-1925.

      Important listing of the published papers of
      scientists, including many engineers. Neglects
      many publications in engineering journals. A
      two-volume **Subject Index** is also available with
      references to kinematics of machinery, steam en-
      gines, and mechanics.

108. Russo, François. **Histoire des sciences et des
      techniques: Bibliogaphie.** Paris: Hermann,
      1954.

      Useful bibliography focusing on European sci-
      ence and technology. Develops the connections
      between science and technology. Includes a list
      of scientific and technical works from the 16th
      to the 20th century.

109. Sarton, George. **A Guide to the History of Sci-
      ence.** New York: Ronald, 1952.

      Created by the founder of the history of sci-
      ence. Includes a significant amount on the his-
      tory of technology, which Sarton saw as part of
      the history of science.

110. Schnitter, N.J. **Bibliography of the History of
      Hydraulic Engineering.** Baden, Switzerland:
      Schnitter, 1979.

      Provides a bibliography of books and papers
      in English, French, Italian and Spanish. Orga-
      nized chronologically.

111. Show, Ralph R. **Engineering Books Available in
      America Prior to 1830.** New York: New York
      Public Library, 1933.

      Useful reference to early books on engineer-
      ing theory. Contains 475 English titles, 188
      French titles, and 29 titles in other languages.

112. Skempton, A.W. **British Civil Engineering, 1640-
        1840: A Bibliography of Contemporary Printed
        Records, Plans, and Books.** London: Mansell,
        1987.

        Useful source for information on civil engi-
        neering.

113. Stapleton, Darwin H., with the assistance of
        Roger L. Shumaker. **The History of Civil En-
        gineering since 1600: An Annotated Bibliog-
        raphy.** New York and London: Garland Publish-
        ing, 1986.

        Contains over 1200 annotated entries dealing
        with civil engineering. Includes chronological
        sections on Theory: Statics, Strength of Materi-
        als and Testing, and sections on Societies, In-
        stitutes, and Education.

114. Technology and Culture. "Current Bibliography in
        the History of Technology." **Technology and
        Culture**, 1964-.

        Published annually in the April issue of
        Technology and Culture. Divided according to
        subject classification and chronological classi-
        fication. No classification for engineering sci-
        ence, but works on the subject can be found under
        general works, technical education, civil engi-
        neering, and energy conversion. Includes an au-
        thor index and a subject index. Edited for many
        years by Jack Goodwin, and since April, 1984
        edited by Stephen H. Cutcliffe, Christine M.
        Roysdon, and Judith A. Mistichelli. Important
        tool.

See also: 623.

CHAPTER II: STUDIES ON THE INTERACTIONS BETWEEN SCIENCE AND TECHNOLOGY

115. Agassi, Joseph. "Between Science and Technology." **Philosophy of Science** 47 (1980): 82-99.

116. _____. "The Confusion between Science and Technology in the Standard Philosophies of Science." **Technology and Culture** 7 (1966): 348-66.

Argues that the differences between science and technology can be seen in their emphasis on corroboration. Views science as a scholarly activity with no real connection to the everyday world and needing no corroboration while technology is concerned with the world and needs corroboration. For a response by Wisdon see item 273.

117. _____. "How Technology Aids and Impedes the Growth of Science." **PSA: Proceedings of the Biennial Meeting of the Philosophy of Science Association.** Volume 2. East Lansing, Mich.: Philosophy of Science Association, 1982, pp. 585-97.

Discusses the role of technology in the development of science.

118. Aitken, Hugh G.J. "Prologue." **Syntony and Spark: The Origins of Radio.** New York: John Wiley & Sons, 1976, pp. 1-30.

Provides a model for understanding the general relationship between science and technology. Focuses on the interactions between scientists, applied scientists, and economic innovators. Develops a model of applied scientists as translators existing between science and technology.

119.   _____. "Science, Technology and Economics: The
        Invention of Radio as a Case Study." **The
        Dynamics of Science and Technology** (item 38),
        pp. 89-111.

        Although focusing on radio, Aitken provides a
        useful model of science-technology interactions.
        Argues that the technological system occupies an
        intermediate position between scientific and eco-
        nomic systems. Concludes that technology trans-
        lates information generated by science into us-
        able economic form and also translates informa-
        tion generated by the economy into a form usable
        by science. Emphasizes the role of individuals
        who can function as translators.

120.   Albring, W. "Das Zusammenwirken von Grundlagen-
        wissen und Technik." **Sitzungsberichte der
        Deutschen Akademie der Wissenschaften zu
        Berlin** 1 (1967): 5-13.

        Discusses the relationship between physics,
        mathematics and technology.

121.   Barnes, Barry. "The Science-Technology Relation-
        ship: A Model and a Query." **Social Studies
        of Science** 12 (February 1982): 166-72.

        Argues that an interactive model in which
        science and technology are equals has replaced
        the hierarchical model in which technology is
        subordinate to science. Concludes that each area
        generates knowledge and each can make use of
        knowledge from the other.

122.   Beer, John J. "The Historical Relations of Sci-
        ence and Technology." **Technology and Culture**
        6 (1965): 547-52.

        Argues that science forms a continuum of gen-
        eralizations from simple empirical formulae to
        sweeping metaphysical theories; also that science
        can be deliberately applied to practical problems
        but can also influence technology through engi-
        neers' unconscious use of these generalizations.

123.   Billington, David P. "Design as Art and Inven-
        tion." **Technology and Science: Important
        Distinctions for Liberal Arts Colleges** (item
        12), pp. 14-26.

Focuses on design as the central difference
between science and engineering. Compares and
contrasts the work of James Watt in designing the
steam engine with that of Thomas Telford in designing bridges. Concludes that art and invention are the central characteristics of design.
Shows that machine advances such as Watt's are
more like science while structural advances such
as Telford's are more like art.

124.   Blenke, Heinz. "Zür Synthese von Wissenschaft
       und Technik." **Mitteilungen der deutschen
       Forschungsgemeinschaft** 4 (1966): 2-26.

       Philosophical study of the synthesis of sci-
ence and technology.

125.   Böhme, Gernot, and Wolfgang van den Daele, and
       Woflgang Krohn. "The 'Scientification' of
       Technology." **The Dynamics of Science and
       Technology** (item 38), pp. 219-50.

       Surveys the relationship between science and
technology from the medieval period through the
19th century. Analyzes how theoretical science
became oriented toward technology and how technology became oriented toward science. Shows how
science developed into special theories of technology while technology developed into theories
of natural structure.

126.   Buchanan, R.A. "The Promethean Revolution: Sci-
       ence, Technology and History." **History of
       Technology** 1 (1976): 73-83.

       Argues that the distinction between science
and technology arose with professionalization
during the mid-19th century. Concludes that to
distinguish science and technology before that
time is a historiographic fallacy.

127.   _____. "The Rise of Scientific Engineering in
       Britain." **British Journal for the History of
       Science** 18 (1985): 218-233.

       Connects scientific engineering to Baconian
natural philosophy. Sees scientific engineering
emerging when engineering acquires a systematic
organization and a theoretical basis. Focuses on
the changing attitudes of engineers. Argues that

recognition by some British universities of an
agreed body of engineering knowledge was crucial.

128.   _____. "Science and Engineering: A Case Study
        in British Experience in the Mid-Nineteenth
        Century." **Notes and Records of the Royal So-
        ciety of London** 32 (1978): 215-223.

        Discusses the role of science and engineering
        in the work of Marc Brunel and his son I.K.
        Brunel.

129.   Bunge, Mario. "Action." **Scientific Research**,
        Volume 2: **The Search for Truth**. New York:
        Springer-Verlag, 1967, pp. 121-150.

        Argues that technological theories are as
        value laden as pure science. Technological theo-
        ries apply either pre-existing scientific theo-
        ries or scientific methods. Discusses the dif-
        ference between scientific laws and technological
        rules. Concludes technological theories are sci-
        entific theories in action. For an earlier ver-
        sion, see item 131.

130.   _____. "Scientific Laws and Rules." **Contempo-
        rary Philosophy; A Survey**, Volume 2: **Philoso-
        phy of Science**. Edited by R. Klibansky.
        Florence: La Nuova Italia Editrice, 1968, pp.
        128-40.

        Argues that technological rules are grounded
        in scientific laws. Distinguishes rules of mod-
        ern technology from rules of art and craft, which
        are grounded in trial and error.

131.   _____. "Technology as Applied Science." **Tech-
        nology and Culture** 7 (1966): 329-47.

        Defines technology as applied science. Ar-
        gues that technology is as theory-laden as pure
        science. Distinguishes between scientific laws
        and technological rules. For a later version of
        the argument see item 129.

132.   Cardwell, Donald S.L. "Science, Technology and
        Industry." **The Ferment of Knowledge: Studies
        in the Historiography of Eighteenth-Century
        Science**. Edited by G.S. Rousseau and Roy
        Porter. Cambridge: Cambridge University
        Press, 1980, pp. 449-83.

Discusses the interaction of science and technology during the Industrial Revolution. Argues that science was a not uncommon factor in the innovative skills of the 18th century. But also argues that technology is not a dependent variable of science. Concludes that science owes more to technology than vice versa.

133. Casimir, H.B.G. "The Relations between Science and Technology." **Storia della fisica del XX secolo**. Edited by C. Weiner. New York: Academic Press, 1977, pp. 447-57.

Discusses the relationships between science and technology during the 20th century.

134. Channell, David F. "The Biological Roots of Nineteenth Century Technology." **Science and Technology, Humanism and Progress: Selections of Papers Presented at the XVIth International Congress of the History of Science**. Bucharest: Academy of the Socialist Republic of Romania, 1982, pp. 59-64.

Argues that George Wilson, first Professor of Technology at the University of Edinburgh, derived his theory of technology from biology. Shows how Wilson saw biology as a way of integrating the observational and transformational sciences that affected technology.

135. _____. "The Distinction between Engineering Science and Natural Science: W.J.M Rankine." **Technology and Science: Important Distinctions for Liberal Arts Colleges** (item 12), pp. 52-59.

Argues that Rankine's 19th-century theory of energetics provides a model of the relationship between his natural science and his engineering science. Shows that his natural science was based on a hypothetical model while his engineering science was based on a phenomenological model set forward in his theory of energetics. Concludes that scientific hypotheses have to be transformed into a phenomenological theory before they can be incorporated into technology.

136.   _____. "Engineering Science as Theory and Prac-
        tice." **Technology and Culture** 29 (1988):
        98-103.

        Response to Fores (see item 154). Shows that
        engineering science involves elements of both
        theory and practice. Proposes that engineering
        science can be analyzed in terms of translations
        between science and technology.

137.   _____. "The Harmony of Theory and Practice:
        The Engineering Science of W.J.M. Rankine."
        Technology and Culture 23 (1982): 39-52.

        Argues that science and technology have inde-
        pendent frameworks and require an intermediate
        mode of knowledge such as engineering science in
        order for there to be any interaction between the
        two. Shows how Rankine created a model of engi-
        neering science which integrated theory and prac-
        tice. Argues that his engineering science de-
        pended on a distinction between laws of actions
        which were theoretical and mathematical, and the
        properties of materials which were practical and
        empirical. Concludes that science can affect
        technology through the laws of nature while tech-
        nology could influence science through the prop-
        erties of materials.

138.   Compton, W. Dale, ed. **The Interaction of Science
        and Technology.** Urbana: University of
        Illinois Press, 1969.

        Contains eight essays which argue against
        Derek Price's claim that there is little interac-
        tion between science and technology. (See item
        233) Most of the authors conclude that science
        does interact with technology through the insti-
        tutionalization of applied science in research
        labs.

139.   Conant, James Bryant. "Induction and Deduction
        in Science and Technology." **Two Modes of
        Thought: My Encounters with Science and Edu-
        cation.** New York: Simon & Shuster, 1964, pp.
        1-31.

        Argues that science is a combination of em-
        pirical inductions and theoretical deductions,
        while the practical arts are empirical-inductive.
        Concludes that the transformation of a practical

art to an applied science involves the introduction of concepts that are developed without re-
gard to their ability to improve practice. Discusses American inventions that required only a
minimum of theory.

140. Condit, Carl W. "Stages in the Relationships be-
        tween Science and Technology." **Technology
        and Culture** 6 (1965): 587-90.

Argues for a dialectical relationship between
science and technology. Believes that the transmission of skills embodied in technical crafts
constitutes a proto-science. Shows that the
emergence of a scientific technology in the 18th
century was prepared by scientific explanations
of the useful properties of materials, the scientific analyses of technical processes and the
technical application of scientific ideas. Comments on Price's paper. See item 233.

141. Constant, Edward W., II. "Communities and Hier-
        archies: Structure in the Practice of Science
        and Technology." **The Nature of Technological
        Knowledge** (item 40), pp. 27-46.

Applies Thomas Kuhn's idea of normal and rev-
olutionary science to technology. Investigates
the similarities and differences between science
and technology. Concludes that scientific knowledge can lead to presumptive anomalies, that is,
the knowledge that a technological system will
fail under some future conditions, and this leads
to technological change. Draws examples from the
development of the turbojet.

142. Cottrell, Allan. "Engineering as a Source of
        Scientific Ideas." **Interdisciplinary Science
        Reviews** 2 (1977): 94-99.

Discusses how engineering influences science.

143. Daniels, George H. "The Pure Science Ideal and
        Democratic Culture." **Science** 156 (1967):
        1699-1705.

Argues that American scientists in the pre-
Civil War period justified research on the basis
of utility while in the post-Civil War period
they turned to science for science's sake and
moved away from contacts with technology.

144. Dimitrieva, M.S., and O.F. Ovcarov.
      "Vzaimodejstvie empiriceskogo i
      teoreticeskogo v texniceskix naukax v
      processe progressivnogo razitija nauki i
      techniki." **Problemy dejatel'nosti ucenogo i
      naucnyx kollektivov** 5 (1973): 102-105.

      Soviet paper entitled "The Interaction of Em-
      pirical and Theoretical Elements in the Techno-
      logical Sciences in the Process of the Progres-
      sive Development of Science and Technology."
      Discusses the role of theoretical knowledge in
      modern technology.

145. Drucker, Peter F. "The Technological Revolution:
      Notes on the Relationship of Technology, Sci-
      ence, and Culture." **Technology and Culture**
      2 (1961): 342-51.

      Argues that the Scientific Revolution had
      little impact on the Technological Revolution and
      shows that a fundamental change in the concept of
      technology was more important. The author be-
      lieves that a central factor in the Technological
      Revolution was the re-ordering of old technolo-
      gies into systematic public disciplines with
      their own conceptual equipment. Concludes that
      technology was a spur to science.

146. Dumas, Maurice. "Rapports entre sciences et
      techniques: Étude générale du point de vue de
      l'histoire des sciences et des techniques."
      Revue de Synthèse 83 (1962): 15-37.

      Discusses the relationship between science
      and technology.

147. Elek, T. "Über die Wechselbeziehungen zwischen
      technischen Wissenschaften, Natur-
      wissenschaften und Mathematik." **Periodica
      polytechnica-Chemical Engineering** 9 (1965):
      237-52.

      Uses dialectical materialism to analyze the
      relationship between engineering science, basic
      science and mathematics.

148. Feibleman, James K. "Pure Science, Applied Science, Technology, Engineering: An Attempt at Definitions." **Technology and Culture** 2 (1961): 305-317.

Argues that all science has application: applied science has the aim of controlling nature while pure science has the aim of understanding nature. Argues for the necessity of an intermediate pursuit between pure and applied science with the goal of getting from theory to practice. Argues that technology is a more practical applied science whose ideal is efficiency while engineering is the most practical activity applying the solution of technology to specific cases; also that the pursuit of practical ends can lead to discovery of abstract principles of science. Concludes that technology shifted from craft to science at the end of the 18th century.

149. Ferguson, Eugene S. "The Mind's Eye: Non Verbal Thought in Technology." **Science** 197 (1977): 827-36.

Analyzes the character of technological knowledge. Concludes that it is more non-verbal than scientific knowledge. Argues that technological knowledge is transmitted in terms of drawings, models, and the direct copying of skills rather than through a written textbook tradition. Important paper.

150. Finch, James Kip. "Engineering and Science: A Historical Review and Appraisal." **Technology and Culture** 2 (1961): 318-332.

Traces the history of the relationship between science and technology from ancient times to the present. Argues that engineering science arose simultaneously with modern science during the 17th and 18th centuries. Concludes that science had little effect on American technology until World War I and the development of the industrial research laboratory.

151. _____. "Rankine: Theory and Practice." **Consulting Engineer** 22 (November 1963): 110-116.

Analyzes the role of theory and practice in the engineering textbooks written by William John Macquorn Rankine during the second half of the 19th century.

152. Fores, Michael. "The History of Technology: An Alternated View." **Technology and Culture** 20 (1979): 853-60.

Attacks the view that technology is applied science and the idea of technology as knowledge. Argues that engineers continue to use trial and error, rather than applying science to technology. Concludes that there are never any best solutions in engineering so that engineering choices cannot be entirely rational or backed by science.

153. _____. "Price, Technology, and the Paper Model." **Technology and Culture** 12 (1971): 621-27.

Critiques the theories of Derek J. de Solla Price on the interaction of science and technology. Argues that new technology derives from old technologies rather than from science.

154. _____. "Transformations and the Myth of 'Engineering Science': Magic in a White Coat." **Technology and Culture** 29 (1988): 62-81.

Attacks the notion of a theoretical approach to technology. Believes that engineering science is the fabrication of 20th century academics. For a response, see items 136, 204.

155. Gibbons, J., and C. Johnson. "The Relationship between Science and Technology." **Nature** 227 (July 11, 1970): 125-27.

Investigates the relationship between science and technology.

156. Greene, John C. "Science and the Practical Arts in Jeffersonian America." **Actes du XIIIe Congrès International d'Histoire des Sciences** 12 (1971, published 1974): 211-17.

Discusses the application of science to technology during the Jeffersonian period.

157. Grove, J.W. "Science as Technology: Aspects of a
      Potent Myth." **Minerva** 18 (1980): 293-312.

      Examines the idea that science and technology
      are indistinguishable.

158. Gruender, C.D. "On Distinguishing Science and
      Technology." **Technology and Culture** 12
      (1971): 456-63.

      Distinguishes science and technology accord-
      ing to scopes and goals. Argues science has a
      general scope with the goal to enlarge knowledge,
      while technology has a narrow scope with a goal
      of profit. Implies technology is applied sci-
      ence.

159. Gutting, Gary. "Paradigms, Revolutions, and
      Technology." **The Nature of Technological
      Knowledge** (item 40), pp. 47-65.

      Argues that science can affect the evaluation
      policy of technological systems as well as the
      systems themselves. Analyzes Mario Bunge's and
      Henryk Skolimowski's debate over whether technol-
      ogy is applied science (see items 129, 131, 253).
      Concludes that technology is neither just applied
      science nor just a set of techniques, but rather
      a body of practical knowledge.

160. Hall, A. Rupert. "Engineering and the Scientific
      Revolution." **Technology and Culture** 2
      (1961): 333-41.

      Shows that 16th and 17th-century philosophers
      and scientists saw the possibility of practical
      applications of science. Argues that technolo-
      gists of the same period had little interest in
      the use of science. Concludes that technology
      had little effect on science except in the area
      of instrumentation.

161. _____. "The Changing Technical Act." **Technol-
      ogy and Culture** 3 (1962): 501-513.

      Concludes that science is necessary for mod-
      ern technology but for its necessary application,
      a certain level of craft skills is necessary as
      well as an industrial framework in which the
      changes affected by science are acceptable.

162.   _____. "On Knowing, and Knowing How To." **History of Technology** 3 (1978): 91-103.

Holds that science contributes to technology in terms of mathematical analysis, methods of experiments, knowledge of relevant laws, and acquaintance with new phenomena. Argues against Layton's claim of knowledge in technology (see item 205). Concludes that rules of technology cannot be independent of science but must be consistent with the laws of science.

163.   _____. "The Scholar and the Craftsman." **Critical Problems in the History of Science.** Edited by Marshall Clagett. Madison: The University of Wisconsin Press, 1959.

Argues that social, intellectual, educational and teleological differences distinguish science and technology. Concludes that knowledge moves from science to technology while techniques and instrumentation move from technology to science.

164.   _____. "What Did the Industrial Revolution in Britain Owe to Science." **Historical Perspectives: Studies in English Thought and Society, in Honour of J.H. Plumb.** Edited by Neil McKendrick. London: Europa, 1974.

Investigates the role of science in the Industrial Revolution.

165.   Holmfeld, John D. "Communication Behavior of Scientists and Engineers." Ph.D. dissertation. Case Western Reserve University, 1970.

Argues that communication between scientists and engineers takes place when the knowledge of one group can be used by members of the other group.

166.   Hudson, L. "Scientific Method and the Nature of Technology." **Nature** 196 (1962): 933-35.

Compares research in science and in technology. Concludes that the scientific method is more important for technology than specific scientific discoveries.

167. Hughes, Thomas P. "The Science-Technology Inter-
         action: The Case of High-Voltage Power
         Transmission Systems." **Technology and Cul-
         ture** 17 (1976): 646-662.

         Focuses on the problem of high-voltage power
      transmission as a case study of the interaction
      of science and technology. Shows the importance
      of institutions, such as the Westinghouse Company
      and the electrical engineering profession, as me-
      diators between individual experiments and an
      emerging engineering theory. Concludes that the
      research methods of engineers could be styled as
      scientific but argues that technological change
      should be seen in the context of expanding tech-
      nological systems rather than focusing only on
      the role of science.

168. Hummon, Norman P. "Organizational Aspects of
         Technological Change." **The Nature of Techno-
         logical Knowledge** (item 40), pp. 67-81.

         Shows that the mix between the art and sci-
      ence of technology can be shifted by new organi-
      zational forms. Focuses on the automobile indus-
      try.

169. Hutchings, Raymond. **Soviet Science, Technology,
         Design: Interaction and Convergence.**
         London: Royal Institute of International Af-
         fairs and Oxford University Press, 1976.

         Argues that invention is made possible by the
      interaction of science, technology and design.
      Shows that the relative prominence of each can
      vary over time. Focuses on how the Russian syn-
      thesis differs from the Western synthesis.

170. IIT Research Institute. **Technology in Retrospect
         and Critical Events in Science (TRACES).** 2
         Volumes. Chicago: Illinois Institute of
         Technology Research Institute, 1968.

         National Science Foundation response to Pro-
      ject Hindsight (see item 250). Concludes that
      major technical advances such as oral contracep-
      tives, electron microscopes and video tape
      recorders were the result of non-mission-oriented

research while advances such as heart pacemakers
and hybrid grains owed as much to non-mission-
oriented research as to mission-oriented
research.

171.   Israel, G., and P. Negrini. "Science and the
        French Revolution." **Scientia** 108 nos. 3-4,
        5-6 (1973): 376-92.

        Outlines the debate which took place over the
role of science during the French Revolution.
Shows how a conception of the relationship between science and technology evolved in which
technology was not subordinated to science.

172.   Janich, Peter. "Physics--Natural Science or
        Technology?" **The Dynamics of Science and
        Technology** (item 36), pp. 3-27.

        Criticizes the self-understanding of physics
by physicists, especially their emphasis on natu-
ralism. Argues that the major parts of physics
that claim to be based on natural science are actually closer to technology. Creates an alter-
nate "constructive" view of physics in which natural science is a consequence of technology
(measurement and observational tools) rather than
technology being seen as an application of science. Concludes that scientists cannot claim
their work is independent of the applications
that are made of it.

173.   Jevons, F.R. "The Interaction of Science and
        Technology Today, Or, Is Science the Mother
        of Invention?" **Technology and Culture** 17
        (1976): 729-38.

        Criticizes the economic model of I.R.C. Byatt
and A.V. Cohen who try to quantify the impact of
scientific discoveries on technology. Argues
that such a model assumes that science discovers
and technology applies. Puts forward an alterna-
tive model in which science supports technological development rather than initiating it. Concludes that technology can affect science just as
science can affect technology.

174. Jobst, Eberhard. "Specific Features of Technol-
        ogy in Its Interrelation with Natural Sci-
        ence." **Contributions to a Philosophy of
        Technology: Studies in the Structure of
        Thinking in the Technological Sciences** (item
        61), pp. 124-33.

        Investigates the relationship between tech-
        nology and science, especially in the area of
        thermodynamics. Conceives of the cooperation be-
        tween science and technology as a dialectical in-
        terrelation. Concludes that the interaction of
        discoveries in both disciplines is necessary for
        the progress of either discipline.

175. _____, and Ulrich Marmai. "Weltanschauliche und
        wissenschaftstheoretische Probleme des Ver-
        haltnisses von Natur- und Technik wis-
        senschaften." **Deutsche Zeitschrift für
        Philosophie** 23, no. 5 (1975): 757-66.

        Focuses on how scientific knowledge is trans-
        ferred into practical technology. Draws examples
        from thermodynamics.

176. Johnston, Ron, and Tom Jagtenberg. "Goal Direc-
        tion of Scientific Research." **The Dynamics
        of Science and Technology** (item 38), pp. 29-
        58.

        Using J.R. Ravetz's historical classification
        of science as pure science, academic science,
        ideologically engaged science, and useful sci-
        ence, argues that most modern science is no
        longer pure but useful.

177. Jonas, Hans. "The Scientific and Technological
        Revolutions: Their History and Meaning."
        Philosophy Today 15, no. 2 (Summer 1971):
        76-101.

        Argues for a metaphysical unity between the
        Scientific Revolution and the Industrial Revolu-
        tion.

178. Kasprzyk, S.F. "On the Concept of Technology and
        its Relation to Science and Technic." **Pro-
        ceedings of the 15th World Congress of Phi-
        losophy** 1 (1973): 321-25.

Characterizes science and technology in terms
of the aims, means and ends of their work.

179. Keller, Alexander. "Has Science Created Technol-
      ogy?" **Minerva** 22 (1984): 160-182.

      Provides a historical survey of the relation-
      ship between science and technology from the time
      of Francis Bacon to the present. Argues that
      science "lays a succession of golden eggs" that
      are exploited by technology.

180. Kline, Ronald. "Science and Engineering Theory
      in the Invention and Development of the In-
      duction Motor, 1880-1900." **Technology and
      Culture** 28 (1987): 283-313.

      Argues that theory in electrical engineering
      differed from the traditional model of engineer-
      ing science in two ways. Shows that theorists
      rank high in electrical engineering and that sci-
      ence of electrical motors did not deal with the
      microscopic world of atoms but with marcroscopic,
      measurable quantities like voltage, current and
      resistance.

181. Klöppel, Kurt. "Die Entwicklung der Ingenieur-
      wissenschaften." **VDI-Zeitschrift** 103, no.
      23 (1961): 1145-53.

      Discusses the development of a science of
      engineering.

182. Konfederatov, I. "O zakonomernostiakh razvitiia
      nauki i teknik na sovremennom etape."
      Voprosy is torii nauki i tekhniki no. 2
      (1970): 40-49.

      Soviet article entitled "On Regularities in
      the Development of Science and Technology on the
      Present-Day State." Refutes Derek Price's views
      on the lack of interaction between science and
      technology. See item 234.

183. Korach, M. "On Methodological Problems of Tech-
      nology." **Periodica polytechnica-Chemical En-
      gineering** 2 (1958): 145-71.

      Discusses the methodological differences be-
      tween science and technology. Characterizes
      technology in terms of the great number of vari-

ables, scale, and cost. Also includes a discussion of model testing and pilot plant experimentation.

184. Kouznetsov, Boris. "Lénine et les principes épistémologiques des prévisions scientifiques, techniques et économiques." **Organon** 7 (1970): 15-36.

Provides a Marxist analysis of the relationship between scientific research and economic-technological progress.

185. Koyré, Alexandre. "Du monde l'à Peu près à l'universe de la Précision." **Critique** 4 (1948): 806-23.

Argues that science and technology are entirely distinct. Believes science is concerned with a world of absolute precision while technology dealt with the world of approximation. Sees technological thought as the result of empirical methods.

186. Krafft, Fritz. "Die Stellung der Technik zür Naturwissenschaft in Antike und Neuzeit." **Technikgeschichte** 37, no. 3 (1970): 189-209.

Contrasts Aristotle's approach to mechanical problems with Galileo's.

187. Kranzberg, Melvin. "The Disunity of Science-Technology." **American Scientist** 56 (Spring 1968): 21-34.

Argues that the interrelationships between science and technology form a spectrum. Sigma Xi Lecture for 1967.

188. _____. "Scientific Research and Technical Innovation." **National Forum** 51 (1981): 27-28.

Argues that technical innovation grew out of old technology, while scientists concern themselves with problems posed by science, not technology.

189. _____. "The Spectrum of Science-Technology." **Journal of the Scientific Laboratories** 48 (December 1967): 47-58.

Argues that the relationships between science
and technology form a spectrum from the lone inventor using little science to the engineering
dependent on science.

190.   _____. "The Unity of Science-Technology."
        American Scientist 55 (March 1967): 48-66.

Reviews 300 years of science and technology.
Argues that social and intellectual needs are
bringing science and technology closer together.
Sigma Xi Lecture for 1966.

191.   _____. "The Wedding of Science and Technology:
        A Very Modern Marriage." **Technology and Sci-
        ence: Important Distinctions for Liberal
        Arts Colleges** (item 12), pp. 27-37.

Proposes the metaphor of a marriage to de-
scribe the symmetrical, non-hierarchical relationship between science and technology. Discusses the historical relationships between science and technology from the time of Plato to the
20th century.

192.   Kuhn, Thomas. "The Relations between History and
        History of Science." **Daedalus** 100 (1971):
        271-304.

Argues that technology absorbs science and
becomes applied science.

193.   Küppers, Günter. "On the Relation between Tech-
        nology and Science--Goals of Knowledge and
        Dynamics of Theories. The Example of Combus-
        tion Technology, Thermodynamics and Fluid-Me-
        chanics." **The Dynamics of Science and Tech-
        nology** (item 38), pp. 113-33.

Argues that because combustion technology,
thermodynamics, and fluid mechanics have a common
theoretical core, their relationship can illumi-
nate the interaction between science and technology. Believes that science seeks insights while
technology builds installations. Assumes that
science and technology are systems of knowledge
influenced by knowledge at hand and by criteria
of relevance. Concludes that science and technology differ in explanatory goals which lead to
different concepts of solution.

194. Langrish, J. "The Changing Relationship between
     Science and Technology." **Nature** 250 (1974):
     614-16.

     Examines the historical changes in the rela-
     tionship between science and technology.

195. Laudan, Rachel. "Cognitive Change in Technology
     and Science." **The Nature of Technological
     Knowledge** (item 40), pp. 83-104.

     Investigates the parallels between problem
     solving in technology and in science. Finds that
     technological problem solving has much in common
     with scientific. Concludes that science can be
     one of the heuristic aids that technologies use
     to solve problems.

196. Layton, Edwin T., Jr., "American Ideologies of
     Science and Engineering." **Technology and
     Culture** 17 (1976): 688-700.

     Examines the ideologies of science and tech-
     nology that developed during the 19th and 20th
     centuries. Discovers different ideologies asso-
     ciated with basic science, engineering science,
     and design. Argues that the scientific ideology
     focused on the application of scientific princi-
     ples and made statements about nature; the ideol-
     ogy of engineering science focused on idealiza-
     tion of machines, beams, and heat engines and
     made statements about devices rather than nature;
     the ideology of design was closer to art and fo-
     cused on practical solutions to engineering prob-
     lems rather than engineering knowledge. Believes
     that the relationship between science and tech-
     nology is similar to the idea of symbiosis. Con-
     cludes that the science-technology relationship
     forms a spectrum from design, which is the most
     practical through engineering science, to basic
     science, which is the most abstract.

197. _____. "Conditions of Technological Develop-
     ment." **Science, Technology and Society: A
     Cross-Disciplinary Perspective.** Edited by
     Ina Spiegel-Rosing and Derek de Solla Price.
     London and Beverly Hills: Sage, 1977 pp. 197-
     222.

Focuses on the role of technology in science policy. Includes a historical section which discusses the interaction of science and technology and the development of the technological sciences. Argues that science and technology have become intermixed. Also discusses the institutionalization of applied science in research laboratories. Concludes that policy-making theories will require more research into the historiography of technology.

198. _____. "Engineering and Science as Distinct Activities." **Technology and Science: Important Distinctions for Liberal Arts College** (item 12), pp. 6-13.

Argues against the idea of technology as applied science. Distinguishes three types of cognitive structures in modern engineering: idealized basic science; less idealized engineering science; and design. Emphasizes the role of design in distinguishing technology from science.

199. _____. "European Origins of the American Engineering Style of the 19th Century." **Scientific Colonialism: A Cross-Cultural Comparison.** Edited by Nathan Reingold and Marc Rothenberg. Washington, D.C.: Smithsonian Institution Press, 1987, pp. 151-66.

Discusses the role of European ideas, including scientific ideas and methods, on the development of American technology.

200. _____. "The Impact of Science on Technology: Implications for Social Control." **Science, Technology, and Society** 16 (February 1980): 1-3.

201. _____. "Millwrights and Engineers, Science, Social Roles, and the Evolution of the Turbine in America." **The Dynamics of Science and Technology** (item 38), pp. 61-87.

Argues against the idea of technology as applied science. Uses the example of the development of the water turbine to attack the dichotomy between scientific and empirical inventions. Shows how the American Francis turbine was a Hegelian synthesis between the approaches of engineers and millwrights. Concludes that mill-

wrights with little knowledge of international
science and engineering used methods that can be
labeled "scientific" to generate needed technological knowledge.

202. _____. "Mirror-Image Twins: The Communities of
Science and Technology in 19th-Century America." **Technology and Culture** 12 (1971): 562-
80.

Argues against the idea of technology as simply applied science. Shows that the communities
of science and technology share many of the same
values in reversed order. Uses examples from the
strength of materials to show that engineering
creates the analogues of experimental and theoretical science. Concludes that information can
flow from technology to science as well as from
science to technology. Important paper.

203. _____. "Science and Engineering Design."
Bridge to the Future (item 28), pp. 173-81.

Distinguishes between a design for bridges
based on a deductive mathematical approach and a
design based on an experimental inductive approach. Argues that engineers began to realize
that science could not "be applied to technology
by a process of simple deduction from the laws of
nature." Concludes that a body of engineering
science was necessary in the development of scientific design.

204. _____. "Science as a Form of Action: The Role
of Engineering Science." **Technology and Culture** 29 (1988): 82-97.

Response to Fores (see item 154). Argues
that there are numerous historical examples of
the existence and development of engineering science. Characterizes engineering science as teleological.

205. _____. "Technology as Knowledge." **Technology
and Culture** 15 (1974): 31-41.

Argues that the view of technology as simply
applied science denies the thought component of
technology. Argues instead for a linking of
technology with knowledge. Uses ideas from

Aristotle, Hugh of Saint Victor and the 20th-century historian of science Alexandre Koyre, to support the idea that technology is a system of thought independent of science. Emphasizes the role of design as the central focus of engineer ing thought. Concludes that knowledge can flow symmetrically between science and technology.

206.   _____. "Technology and Science, or 'Vive la petite difference.'" **PSA 76: Proceedings of the 1976 Biennial Meeting of the Philosophy of Science Association.** Volume 2. East Lansing, Mich.: Philosophy of Science Association, 1977, pp. 173-84.

Argues that science emphasizes knowing and tends toward generalization and simplification while technology emphasizes doing and tends to account for imperfections of materials.

207.   _____. "Through the Looking Glass, or News from Lake Mirror Image." **Technology and Culture** 28 (1987): 594-607.

Presidential Address to the Society for the History of Technology. Useful survey of the is-sues surrounding the nature of technological knowledge, such as the role of science and tech-nology in engineering science and engineering de-sign. Contains a useful bibliography.

208.   Leniham, John M.A. "The Triumph of Technology." **Philosophical Journal: Transactions of the Royal Philosophical Society of Glasgow** 6, no. 1 (1969): 12-18.

Examines the career of Lord Kelvin by focus-ing on the relationship between science and tech-nology. Concludes that technology was the inspi-ration for his science.

209.   Lenk, Hans. **Pragmatische Philosophie; Pladoyers und Beispiele für eine praxisnache Philoso-phie und Wissenschaftstheorie.** Hamburg: Hoffmann & Campe, 1975

Contains a discussion on whether technology is applied science. Refers to the works of Mario Bunge, Henryk Skolimowski, Friedrich Rapp and others.

210. _____. "Zu neueren Ansatzen der Technikphiloso-
     phie." **Techne, Technik, Technologie:
     Philosophische Perspektiven.** Edited by H.
     Lenk and S. Moser. Munich: Pullach, 1973,
     pp. 198-231.

     Discusses the methodological distinctions be-
     tween science and technology. Also analyzes the
     differences within the technological sciences.

211. Ludewig, Walter. "Die Ingenieursarbeit--eine
     Kunst oder eine Wissenschaft?" **Fridericiana;
     Zeitschrift der Universität Karlsruhe** 1, no.
     1 (1967): 23-27.

     Raises the question whether engineering is an
     art or a science.

212. McDermott, John J. "Nature and Human Artifact."
     **Proceedings of the XVth World Congress of
     Philosophy, Varna, Bulgaria.** Volume 3.
     Sofia: Sofia Press Production Center, 1974,
     pp. 119-22.

     Calls for an end to the dualism between na-
     ture and artifact. Argues for accepting the
     built world as the human's place.

213. McKendrick, Neil. "The Role of Science in the
     Industrial Revolution: A Study of Josiah
     Wedgwood as a Scientist and Industrial
     Chemist." **Changing Perspectives in the His-
     tory of Science.** Edited by Mikulas Teich and
     Robert Young. Dordrecht and Boston: D.
     Reidel, 1973, pp. 279-318.

     Argues that scientific knowledge was largely
     decentralized and amateur during the 18th century
     but was connected to a Baconian ideology of mate-
     rial improvements. Concludes that such an eco-
     nomically motivated science was used by Josiah
     Wedgwood in his pottery manufacturing and also
     underlay the Industrial Revolution.

214. McKeon, R. "A Study of the History of 19th-Cen-
     tury Science and Technology: Engineering
     Science in the Works of Navier." **Actes XIIIe
     Congrès International d'Histoire des Sciences**
     11 (1974): 321-27.

Discusses Navier's work in terms of engineering science.

215. Makeeva, V.N. "V.I. Lenin i nauchnoteckhnicheskii otdel VSNKh." **Voprosy istorii estestvoznaniia i tekhniki** no. 1 (1970): 46-51.

Soviet article entitled "V.I. Lenin and the Department of Science and Technology of the Council of National Economy." Describes Lenin's ideas on the relationship between science and industrial technology.

216. Mantell, M.I. "Scientific Method--A Triad." **Contributions to a Philosophy of Technology: Studies in the Structure of Thinking in the Technological Sciences** (item 61), pp. 115-123.

Puts forward three patterns of the scientific method -- basic research, applied research, and systems approach. Argues that engineers use the last two more than the first. Concludes that applied research differs from basic in terms of the types of assumptions that are used. Shows that assumptions and standards in engineering involve value judgments requiring compromises between safety, cost, durability, reliability, efficiency, convenience, and aesthetics.

217. Mathias, Peter. "Who Unbound Prometheus? Science and Technological Change, 1600-1800." **Science and Society, 1600-1900.** Edited by Peter Mathias. Cambridge: Cambridge University Press, 1972, pp. 54-80.

Investigates the importance of scientific knowledge in the development of industrial growth relative to other factors.

218. Mayr, Otto. "The Science-Technology Relationship as a Historiographic Problem." **Technology and Culture** 17 (1976): 663-72.

Argues that historians have proposed several models and approaches for understanding the relationship between science and technology including: science and technology as opposites; science and technology in a hierarchical relation; science and technology as points on a spectrum; and

science and technology as poles of a magnet.
Concludes that the historian's task is not to analyze the relationship between science and technology but to analyze what other eras and cultures have thought it to be.

219. Meletschenko, IU.S., and S.V. Shukhardin. "V.I.
      Lenin i nekotorye problemy tekhniki."
      Voprosy istorii estestvoznaniia i tekhniki
      no. 1 (1970): 20-28.

      Soviet article entitled "V.I. Lenin and the
      Problem of Technology." Discusses the role of
      technology in the modern revolution in science.

220. Molella, Arthur P., and Nathan Reingold.
      "Theorists and Ingenious Mechanics: Joseph
      Henry Defines Science." **Science Studies** 3
      (1973): 323-351.

      Discusses Joseph Henry's views on the rela-
      tionship between science and technology.

221. Moraru, I. "Quelques particularités de la con-
      naissance dans les sciences techniques: Pre-
      misses pour une épistémologie des sciences
      techniques." **Proceedings of the 15th World
      Congress of Philosophy** 2 (1973): 343-46.

      Discusses engineering science in terms of
      methods and subject matter.

222. Moser S. "Metaphysics of Technology." **Philoso-
      phy Today** 9 (1971): 129-56.

      Compares the nature of scientific and techno-
      logical experiments.

223. Mulkay, Michael. "Knowledge and Utility: Impli-
      cations for the Sociology of Knowledge." **So-
      cial Studies of Science** 9 (February 1979):
      63-80.

      Argues for an interactive model of the rela-
      tionship between science and technology. Con-
      cludes that his model will influence the rela-
      tionship of science to society.

224. Muller, J. "Zür Bestimmung der Begriffe
      'Technik' und 'technisches Gesets.'"
      Deutsche Zeitschrift für Philosophie 15
      (1967): 1431-49.

      Analyzes the laws of technology and their re-
      lation to the natural sciences.

225. _____. "Zum Verhaltnis von Naturwissenschaft
      und Technik." **Freiberger Forschungshefte** D
      53 (1967): 163-70.

      Analyzes the similarities and differences be-
      tween science and technology. Concludes that
      they apply the same methods but vary in their
      aims.

226. Multhauf, Robert. "The Scientist and the
      'Improver' of Technology." **Technology and
      Culture** 1 (1959): 38-47.

      Argues that traditional distinctions between
      science and technology as knowledge and utility
      do not hold up under close scrutiny. Also be-
      lieves that the distinction between discoveries
      versus inventions is less reliable. But he con-
      cludes that the actual scientist and the improver
      of technology seem distinct.

227. Mumford, Lewis. "Science as Technology." **Pro-
      ceedings of the American Philosophical Soci-
      ety** 105, no. 5 (October 1961): 506-511.

      Argues against the Baconian ideal of linking
      science to technology.

228. Munch, Richard. "Modern Science and Technology--
      Differentiation of Interpretation." **Interna-
      tional Journal of Comparative Sociology** 24
      (September-December 1983): 157-75.

      Provides a sociological analysis of the rela-
      tionship between science and technology.

229. Musson, A.E., and Eric Robinson. **Science and
      Technology in the Industrial Revolution.**
      Manchester: Manchester University Press,
      1969.

Contains a series of separate essays. Argues that science played an important role in the Industrial Revolution. Focuses on chemistry but also deals with steam power.

230. Nakagawa, Yasuo. "Relationships between 'Properties of Matter' in Physics Textbooks in the Late 19th Century and Technical Education in the Industrial Revolution." **Kagakusi Kenkyu** 16 (1977): 161-66.

In Japanese.

231. Odqvist, Folke K.G. "The Connections between Science and Engineering." **Lychnos** (1937): 148-60.

In Swedish with an English summary.

232. Price, Derek J. de Solla. **The Difference between Science and Technology**. Detroit: Thomas Alva Edison Foundation, 1968.

Investigates the relationship between science and technology. Emphasizes that there are more differences than similarities between science and technology.

233. _____. "Is Technology Historically Independent of Science? A Study in Statistical Historiography." **Technology and Culture** 6 (1965): 553-68.

Argues that the difference between science and technology is one of communication; the scientist publishes and the technologist does not. Implies that the interaction between science and technology occurs during training. Believes that each type of professional uses only the state of the art of the other field that is learned in school. For a response, see items 138 and 140.

234. _____. "Issledovanie o' issledovanii." **Voprosy istorii estestvoznaniia i tekhniki** no. 2 (1970): 30-39.

Article in Russian entitled "Study about Study." Includes the argument that there is little important interaction between science and technology.

235.   _____. "Notes Towards a Philosophy of the Sci-
        ence/Technology Interaction." **The Nature of
        Technological Knowledge** (item 40), pp. 105-
        114.

        Argues that technology influences science
        through the craft of experimental science. Shows
        that this craft tradition does not test new hy-
        potheses, but provides new information through
        "artificial revelation" which affects what sci-
        ence must explain. Takes his examples from
        Galileo's use of the telescope and Volta's use of
        the Voltaic pile.

236.   _____. "The Structure of Publication in Science
        and Technology." **Factors in the Transfer of
        Technology**. Edited by W.H. Gruber and D.G.
        Marquis. Cambridge, Mass.: M.I.T. Press,
        1969, pp. 91-104.

        Argues that technology is research where the
        main product is a machine, product or process
        rather than a paper as it is in science. Con-
        cludes that technology has a formal structure
        identical to science.

237.   Pugsley, A.G. "Scientific Discovery." **The Engi-
        neer Centenary Number** (1956): 163-65.

        Examines the influence of science on engi-
        neering. Includes a discussion of friction, arch
        construction, and the development of the electric
        strain gauges.

238.   Rabkin, Yakov M. "Science and Technology: Can
        One Hope to Find a Measurable Relationship?"
        Fundamenta Scientiae 3-4 (1981): 420-21.

        Argues for a symmetrical relationship between
        science and technology rather than a hierarchical
        one.

239.   Rapp, Friedrich. **Analytical Philosophy of Tech-
        nology**. Translated by Stanley R. Carpenter
        and Theodore Langenbruch. Dordrecht,
        Holland: Reidel, 1981.

        Important book on the philosophy of technol-
        ogy. Includes a discussion of the relationship
        between science and technology. Includes a sec-

tion on the natural sciences and the engineering
sciences.

240. _____. "Technology and Natural Science--A
    Methodological Investigation." **Contributions
    to a Philosophy of Technology: Studies in
    the Structure of Thinking in the Technologi-
    cal Sciences** (item 61), pp. 93-114.

    Distinguishes science and technology by means
    of their respective aims and procedures. Shows
    that technical problems lead to further scien-
    tific research. Concludes that the direction of
    the natural sciences is determined by technical
    problems.

241. Regan, M.D. "Basic and Applied Research: A Mean-
    ingful Distinction?" **Science** 155 (1967):
    1383-86.

    Argues that neither goals nor applicability
    is useful in distinguishing basic and applied re-
    search. Concludes that production of new knowl-
    edge and use of existing knowledge are more use-
    ful distinctions.

242. Reingold, Nathan, and Arthur Molella, eds. "The
    Interaction of Science and Technology in the
    Industrial Age." **Technology and Culture** 17
    (1976): 621-724.

    Proceedings of the Burndy Library Conference.
    Contains items 167, 173, 196, 218, 563, 1176.

243. Reiss, Howard. "Human Factors at the Science-
    Technology Interface." **Factors in the Trans-
    fer of Technology.** Edited by W.H. Gruber and
    D.G. Marquis. Cambridge, Mass.: M.I.T.
    Press, 1969, pp. 105-116.

    Distinguishes between science and applied
    science. Argues that they differ on the basis of
    their activities and subject matters.

244. Rosenberg, Nathan. "The Growing Role of Science
    in the Innovation Process." **Science, Tech-
    nology, and Society in the Time of Alfred
    Nobel.** Oxford: Pergamon Press, 1982.

245.   Rumpf, H.   "Gedanken zur Wissenschaftstheorie der
        Technikwissenschaften."   **Techne, Technik,**
        Technologie: Philosophische Perspektiven.
        Edited by H. Lenk and S. Moser.   Munich:
        Pullach, 1973, pp. 82-107.

        Analyzes science and technology in terms of
        aims, methods, subject matter and structure of
        their propositions.   Concludes that there are
        more similarities than differences.

246.   Saunders, O.A.   "The Scientist's Contribution to
        Mechanical Engineering."   **The Engineer** 210
        (1960): 760-61.

        Argues that science contributes new princi-
        ples, new materials and production processes, and
        new designs to mechanical engineering.

247.   Schofield, Robert E.   "On the Equilibrium of a
        Heterogeneous Social System."   **Technology and**
        Culture 6 (1965): 591-95.

        Provides a critique of Price's "thermodynamic"
        model of the relationship between science and
        technology.   See item 233.

248.   Sebestik, Jan.   "The Rise of Technological Sci-
        ence."   **History and Technology** 1 (1983): 25-
        44.

        Discusses the development of engineering sci-
        ence.

249.   Semenev, G.I.   "O prirode i specifike
        texniceskogo znanija."   **Naucnye doklady**
        vyssej skoly: Filosofskie nauki 6 (1969):
        40-46.

        Article entitled "About the Nature and the
        Specific Character of Technological Knowledge."
        Argues that some concepts describe natural phe-
        nomena while other concepts refer to how technol-
        ogy makes use of natural phenomena.

250.   Sherwin, C.W., and R.S. Isenson.   "Project Hind-
        sight."   **Science** 156 (1967): 1571-77.

        Discusses a study of the relationship between
        science and technology conducted by the Depart-
        ment of Defense.   Concludes that applied research

was more important in producing new weapon systems than basic research. See item 170 for a response by the National Science Foundation.

251. Simon, Herbert. **The Sciences of the Artificial.**
      Cambridge, Mass.: M.I.T. Press, 1969.

      Categorizes artificial phenomena as having an
      air of contingency while natural phenomena have
      an air of necessity. Argues for the creation of
      a science of design or a science of the artifi-
      cial. Important work.

252. Skolimowski, Henryk. "On the Concept of Truth in
      Science and Technology." **Proceedings of the
      XIVth International Congress of Philosophy.**
      Volume 2. Vienna: Herder, 1968, pp. 533-59.

      Argues that science aims at enlarging knowl-
      edge through better theories while technology
      aims at creating artifacts through increasing ef-
      fectiveness. Concludes that the aims and means
      of science and technology are different.

253. _____. "The Structure of Thinking in Technol-
      ogy." **Technology and Culture** 7 (1966): 371-
      83.

      Distinguishes science and technology accord-
      ing to different types of thinking. Argues that
      while science investigates, technology creates
      and therefore uses a different mode of thought.
      Shows that surveyors think in terms of geodesics,
      civil engineers think in terms of durability and
      mechanical engineers think in terms of effi-
      ciency.

254. Smith, Crosbie. "Engineering the Universe:
      William Thomson and Fleeming Jenkin on the
      Nature of Matter." **Annals of Science** 37
      (1980): 387-412.

      Analyzes how ideas from engineering influ-
      enced theories of matter during the second half
      of the nineteenth century.

255. Smith, Cyril Stanley. "Art, Technology, and Sci-
      ence: Notes on Their Historical Interac-
      tion." **Technology and Culture** 11 (1970):
      493-549.

Argues that it is misleading to divide human
actions into art, science or technology. Shows
that the engineer is both scientist and artist.
Emphasizes a materials-oriented view of history.
Traces examples from the ancient past to the present. Also includes a discussion of the relationship between materials design and structural
engineering in mechanical technology.

256.   Stapleton, Darwin H., and Edward C. Carter, II.
       "'I Have the Itch of Botany, of Chemistry, of
       Mathematics ... strong upon me': **The Science
       of Benjamin Henry Latrobe.**" Proceedings of
       the American Philosophical Society** 128
       (1984): 173-92.

       Includes a discussion of Latrobe's idea of a
science of engineering.

257.   Steffens, Henry John, and H.N. Muller, eds. **Sci-
       ence, Technology and Culture.** New York: AMS,
       1974.

       Presents the results of a series of meetings
between faculty at the University of Vermont and
executives at Western Electric. Contains an article by L. Pearce Williams on the historical
distinctions between science and technology.

258.   Thackray, Arnold. "The Industrial Revolution and
       the Image of Science." **Science and Values:
       Patterns of Tradition and Change.** Edited by
       Arnold Thackray and Everett Mendelsohn. New
       York: Humanities Press, 1974, pp. 3-18.

       Discusses how the Industrial Revolution
shaped a new image of science.

259.   Tondl, L. "On the Concepts of 'Technology' and
       'Technological Sciences.'" **Contributions to
       a Philosophy of Technology: Studies in the
       Structure of Thinking in the Technological
       Sciences** (item 61), pp. 1-18.

       Argues that the conception of technological
sciences as merely applied sciences is not ten-
able. Believes that the technological sciences
have a cognitive function. Concludes that the
technological sciences do not restrict themselves
to the imitation of nature but create a new
nature.

260. Ubbelohde, A.R.   "Edwardian Science and Technol-
      ogy: Their Interactions." **British Journal
      for the History of Science** 1 (3) (1963):
      217-26.

      Discusses the interaction of science and
      technology in early 20th-century Britain.

261. Usher, Abbott Payson.   "The Industrialization of
      Modern Britain." **Technology and Culture** 1
      (1960): 109-127.

      Includes a discussion of the role of science
      and mathematics in the industrialization of
      Britain. Argues for a concept of evolutionary
      change.

262. Vincenti, Walter G.   "Control-Volume Analysis: A
      Difference in Thinking between Engineering
      and Physics." **Technology and Culture** 23
      (1982): 145-74.

      Argues that the differences in thinking be-
      tween engineering and physics arise from the dif-
      ferences in physical problems and economic con-
      straints in each field. Defines the concept of
      engineering science. Shows that engineering sci-
      ence is similar to science in that it uses the
      same natural laws, that it is diffused through
      textbooks, articles, and the classroom, and that
      it is cumulative. Concludes that engineering
      science is different from science in that the
      purpose of engineering science is the production
      of an artifact while the purpose of science is
      the production of knowledge. Important article.

263. _____.   "Technological Knowledge without Sci-
      ence: The Innovation of Flush Riveting in
      American Airplanes, ca.1930-ca.1950." **Tech-
      nology and Culture** 25 (1984): 540-576.

      Argues that the introduction of flush rivet-
      ing in the American airplane industry did not in-
      volve any dependence upon science. Shows that
      the process reveals a different kind of techno-
      logical knowledge. Distinguishes between de-
      scriptive knowledge, (knowledge of fact) and pre-
      scriptive knowledge (knowledge of procedure or
      operation). Finds that descriptive knowledge is
      closer to science since it cannot be willfully

adjusted by technologists the way prescriptive
knowledge can. Also argues for a tacit knowledge
(implicit, wordless, pictureless) that is also
procedural. Emphasizes the role of design.

264. Volosevic, O.M., and Ju.S. Melescenko.
     "Texniceskie nauki i ix mesto v sisteme
     naucnogo znanija." **Metodologiceskie problemy
     vzaimosvjazi i vzaimodejstvija nauk.** Edited
     by M.V Mostepanenko. Leningrad, 1970, pp.
     242-62.

     Soviet article entitled "The Technological
     Sciences and Their Place within the System of
     Scientific Knowledge." Analyzes the relationship
     between science and technology. Discusses the
     classification of the technological sciences.

265. Walentynowicz, Bohdan. "On Methodology of Engi-
     neering Design." **Proceedings of the XIVth
     International Congress of Philosophy.** Volume
     2. Vienna: Herder, 1968, pp. 587-90.

     Argues that scientific cognition is based on
     analysis and induction while engineering cogni-
     tion is based on synthesis and creativity. Con-
     cludes that a scientific methodology has a lim-
     ited use in engineering design. Argues that a
     methodology of engineering design should be based
     on general systems theory, operations research,
     game theory and cybernetics.

266. Wartofsky, Marx W. **Conceptual Foundations of
     Scientific Thought.** New York: Macmillan,
     1968.

     Argues that technical rules as prescriptive
     formulations of laws contain the germinal form of
     a developed science.

267. Watson-Watt, Sir Robert. "Technology in the Mod-
     ern World." **Technology and Culture** 3
     (1962): 385-93.

     Discusses the relationship between science
     and technology in the modern world. Argues that
     science and technology have differentiated iden-
     tities but share a common blood-stream like
     Siamese twins. Concludes they differ in outlook.

Sees science as a representational art while
technology adapts the processes and materials
identified by science to an end.

268. Weingart, Peter. "The Relation between Science
      and Technology--A Sociological Explanation."
      The Dynamics of Science and Technology (item
      38), pp. 251-86.

      Provides a comparative analysis of the pro-
      duction of knowledge in science and technology.
      Argues that the introduction of modern science
      differentiates theoretical and practical concerns
      of knowledge production. Concludes that this
      process leads to the scientification of practical
      modes of knowledge production.

269. _____. "The Structure of Technological Change:
      Reflections on a Sociological Analysis of
      Technology." **The Nature of Technological
      Knowledge** (item 40), pp. 115-42.

      Develops a sociological theory of technologi-
      cal change in the 19th-century universities. Re-
      lates technological knowledge to different types
      of institutionalized orientations such as univer-
      sities and professional societies. Discusses the
      commonalities that exist between science and tech-
      nology.

270. Wendt, H. "Die Wechselbeziehungen zwischen tech-
      nischen und theoretischen Wissenschaften."
      **Wissenschaftliche Zeitschrift der Technischen
      Hochschule Karl-Marx-Stadt** 11 (1969): 192-
      226.

      Analyzes the cognition and application of
      technological laws.

271. Wisdom, J.O. "The Need for Corroboration; Com-
      ments on Agassi's Paper." **Technology and
      Culture** 7 (1966): 367-70.

      Argues that applied science in concerned with
      understanding and extending knowledge while tech-
      nology is concerned with using knowledge. Con-
      cludes that applied science is a step toward do-
      ing, but it is wrong to link it with technology
      rather than pure science. Response to a paper by
      Agassi, see item 116.

272. Wise, M. Norton, and Crosbie Smith.
      "Measurement, Work and Industry in Lord
      Kelvin's Britain." **Historical Studies in the
      Physical and Biological Sciences** 17 (1986):
      147-73.

      Investigates the relationship between mea-
      surement and industry in the work of Lord Kelvin.
      Demonstrates how the image of the steam engine
      provided a model for Kelvin's use of precision
      measurement in physics. Shows how scientific
      measurement was linked to issues of values. Con-
      cludes that Kelvin's science was a form of fac-
      tory physics.

273. Wojick, David. "The Structure of Technological
      Revolutions." **The History and Philosophy of
      Technology.** Edited by G. Bugliarello and
      D.B. Doner. Urbana: University of Illinois
      Press, 1979.

      Focuses on the evaluation policy of techno-
      logical systems. Provides a model in which sci-
      ence can affect technology through changes in the
      procedure of evaluation.

274. Woolf, Harry. "Basic Research and Industrial En-
      terprise." **Minerva** 22 (1984): 183-95.

      Analyzes the relationship between academic
      research and technological innovations. Includes
      an analysis of the role of Lord Kelvin.

275. Zvorikine, A. "The History of Technology as a
      Science and as a Branch of Learning: A So-
      viet View." **Technology and Culture** 2
      (1961): 1-4.

      Argues that technology is bound directly to
      the laws of natural science while social and eco-
      nomic activity are bound to natural science
      through technology. Concludes that the history
      of technology is a science.

276. _____. "Technology and the Laws of Its Develop-
      ment." **Technology and Culture** 3 (1962):
      443-58.

Includes a discussion of the relationship between science and technology as dialectical. Using Lenin, the author argues that solving a technological problem based on known scientific laws
may lead to new discoveries of the properties of
things and advance science. Argues that science
shows the possible variants of solving a technical problem but does not determine the exact solution.

See also: 5, 7, 9, 10, 32, 55, 74, 425, 478, 674, 974,
990, 1168, 1171, 1174, 1176, 1177, 1182, 1209, 1220,
1223, 1224, 1337, 1351, 1352, 1357, 1358, 1470, 1471,
1491, 1492.

CHAPTER III: BIOGRAPHICAL STUDIES AND COLLECTED WORKS

BIOGRAPHICAL DICTIONARIES

277.   Académie des Sciences. **Index Biographique des
       membres et correspondants de l'Académie des
       Sciences 1666-1954.** Paris, 1954.

       Provides biographical details of members of
       the Academie of Sciences, which included many who
       contributed to the creation of engineering sci-
       ence.

278.   Barr, E. Scott. **An Index to Biographical Frag-
       ments in Unspecialized Scientific Journals.**
       University, Alabama: The University of
       Alabama Press, 1973.

       Lists biographical information, usually obit-
       uaries, in 19th- and 20th-century scientific
       journals. Includes both engineers and scien-
       tists.

279.   Bell, Samuel Peter, ed. **A Biographical Index of
       British Engineers in the 19th Century.** New
       York and London: Garland Publishing, 1975.

       Presents an index of biographical sketches of
       3,500 British engineers. Provides the birth,
       death dates and the journal in which the sketch
       appeared. Identifies the field of each engi-
       neer.

280.   Blanchard, Anne. **Dictionnaire des ingénieurs
       militaires, 1691-1791** (Collection du Centre
       d'Histoire Militaire et d'Etudes de Défense
       Nationale, 14). Montpellier: Université
       Paul-Valéry, 1981.

       Provides biographical details of over 1000
       engineers.

281. Gillispie, Charles C., ed. **Dictionary of Scien-
      tific Biography.** 16 Volumes. New York:
      Scribner's, 1970-1980.

      Contains signed biographical articles, with
      bibliographies, of scientists and engineers. Em-
      phasizes science over technology, but engineers
      with a scientific interest are represented.

282. Lee, Antoinette, under the direction of Harold
      Skramstad. **A Biographical Dictionary of
      American Civil Engineers.** New York: American
      Society of Civil Engineers, 1972.

      Presents brief biographies of almost 200 en-
      gineers born after the Civil War. Includes in-
      formation on education, career and publications.

283. Leprince-Ringuet, Louis, ed. **Les inventeurs
      célèbres. Sciences physiques et applica-
      tions.** Paris: Mazenod, 1950.

      Contains biographical sketches of engineers
      and scientists.

284. Poggendorf, Johann Christian. **Biographisch-
      literarisches Handworter-buch zür Geschichte
      der exakten Wissenschaften.** 6 volumes.
      Leipzig, 1863-1936.

      Provides biographical information and a list
      of publications for a great number of 19th- and
      20th-century scientists and engineers. Important
      research tool.

285. Royal Society of Edinburgh. **Index of Fellows of
      the Royal Society of Edinburgh Elected Novem-
      ber 1783-July 1883.** Compiled by Flora
      Bennet, Hugh Frew, and Maneene Melrose.
      Edited by Hugh Frew. Edinburgh: Scotland's
      Cultural Heritage, 1984.

      Includes biographies of several persons who
      contributed to engineering science.

286. Roysdon, Christine, and Linda A. Khatri. **Ameri-
      can Engineers of the Nineteenth Century: A
      Biographical Index.** New York and London:
      Garland Publishing, 1978.

Provides biographical information on 2,200
nineteenth-century American engineers. Lists
birth and death dates, field of engineering and
references to biographical notices.

BIOGRAPHIES AND AUTOBIOGRAPHIES

287. Allan, D.G.C. **William Shipley, Founder of the
      Royal Society of Arts.** Foreward by His Royal
      Highness the Prince Philip, Duke of
      Edinburgh, President of the Royal Society of
      Arts. London: Hutchinson & Co., 1968.

      Provides a biography of the founder of the
      Society for the Encouragement of Arts, Manufac-
      tures, and Commerce, later known as the Royal
      Society of Arts. Discusses the role of science
      in the technical arts.

288. Allard, Michel. **Henri-Louis Duhamel du Monceau
      et le ministère de la marine.** Montreal:
      Leméac, 1970.

      Discusses the career of one of the signifi-
      cant pioneers in 18th-century naval architecture.

289. Allen, Jack. "The Life and Work of Osborne
      Reynolds." **Osborne Reynolds and Engineering
      Science Today.** Edited by D.M. McDowell and
      J.D. Jackson. Manchester: Manchester Univer-
      sity Press, 1970.

      Presents a useful biographical study of a
      significant figure in the development of engi-
      neering science.

290. Arago, Dominique François Jean. **Historical Eloge
      of James Watt.** New York: Arno, 1975.

      Reprint of the 1839 edition. Provides useful
      details concerning the life of James Watt and the
      role of science in his inventions.

291. Artobolevskii, Ivan Ivanovich, and Aleksei
      Nikolaevich Bogliubov. **Leonid Valdimirovich
      Assur (1878-1920).** Moscow: Nauka, 1971.

Provides details on the life of L.V. Assur
and his important contributions to mechanical engineering and theoretical mechanics.

292.   Aubry, P.-V. **Monge, le savant ami de de
       Napoleon.** Paris, 1954.

Provides a biography of Gaspard Monge, one of
the founders of the École Polytechnique.

293.   Beckett, Derrick. **Brunel's Britain.** North
       Pomfret, Vt.: David & Charles, 1980.

A study of the interaction of architecture
and engineering in the work of Isambard Kingdom
Brunel. Contains a discussion of Brunel's use of
the strength of materials, the theory of stability, and the statics of suspension bridges. Includes little on his work on railways and steam-
boats.

294.   _____. **Stephenson's Britain.** North Pomfret,
       Vt.: David & Charles, 1984.

Describes the engineering works of George and
Robert Stephenson. Analysis of their bridges is
more satisfactory than the biographical details
of their lives. Appendix contains a discussion
of the testing of components for the Britannia
Bridge.

295.   Berkel, Klaas van. **Isaac Beekman (1588-1637) en
       de mechanisering van het wereldbeeld.**
       Amsterdam: Rodopi, 1983.

Discusses the life of the 17th-century Dutch
natural philosopher Isaac Beekman and analyzes
his contributions to the mechanization of the
world picture. Shows that Beekman concluded that
the application of science to technology was necessary for improvement of the world. Author provides an English summary.

296.   Birkzwager, J.M. **Dr. B.J. Tideman, 1834-1883.**
       Leiden: E.J. Brill, 1970.

Provides a study of the contributions of B.J.
Tideman, chief engineer of the Dutch navy and
lecturer in shipbuilding at the Polytechnical

School at Delft, to the modernization of ship
building. Discusses his use of models and experiments.

297. Bishop, Philip W. "John Ericsson (1803-89) in
England." **Transactions of the Newcomen Society** 48 (1976-77): 41-52.

Discusses the relationship between Ericsson's
stay in England and his later inventions.

298. Bishop, R.E.D. "Alexander Kennedy: The Elegant
Innovator." **Transactions of the Newcomen Society** 47 (1974-76): 1-8.

Discusses the life of Alexander Kennedy, who
was professor of engineering at University College, London and who established Britain's first
engineering laboratory.

299. Bogoliubov, Alekséi. **Un héroe español del pro-
greso: Agustin de Betancourt.** Madrid: Semi-
narios y Ediciones, 1973.

Presents a biography of the Spanish-born engineer Agustin de Betancourt, who worked on several technological projects for Alexander I of
Russia. Discusses his early experiments in steam
power and his important theoretical work on machines. Connects Betancourt's theoretical work
with the need, brought on by the Industrial Revolution, to systematize knowledge about machines.

300. Booker, P.J. "Gaspard Monge (1746-1818) and His
Effect on Engineering Drawing and Technical
Education." **Transactions of the Newcomen So-
ciety** 34 (1961-62): 15-36.

Describes Monge's use of descriptive geometry
in solving engineering problems. Describes his
role in the creation of the engineering curriculum at the École Polytechnique. Includes an outline of his course in descriptive geometry.

301. Booth, L.G. "Thomas Tredgold (1788-1829): Some
Aspects of His Work." **Transactions of the
Newcomen Society** 51 (1979-80): 57-94.

Discusses the work of an important pioneer in
materials testing and theory of construction.

302. Bottema, O. "F. Reuleaux, Filosoof der Tech-
      niek." **Algemeen Nederlands Tijdschrift voor
      Wijsbegeerte en Psychologie** 59, no. 1 (April
      1967): 28-32.

      Provides a study of Franz Reuleaux's attempt
      to develop a theory of machines in the 19th cen-
      tury.

303. Boucher, Cyril T.G. "John Rennie (1761-1821)."
      Transactions of the Newcomen Society 34
      (1961-62): 1-14.

      Provides the biographical details of Rennie's
      life and the relationship between his knowledge
      of structures and his bridges.

304. _____. **John Rennie, 1761-1821.** Manchester:
      Manchester University Press, 1963.

      Presents a biography of a major civil engi-
      neer. Argues that one of his greatest accom-
      plishments was the application of theory to prac-
      tice, especially the theory of structures.

305. Bracegirdle, Brian, and Patricia H. Miles.
      Thomas Telford. Newton Abbot, Devon: David &
      Charles, 1973.

      Provides an account of Telford's life.

306. Bradley, M. "Scientific Biography of Gaspard de
      Prony." Ph.D. dissertation. Lancaster Poly-
      technic (UK), 1984.

      Biography of a leading contributor to French
      engineering science and one of the founders of
      the École Polytechnique.

307. Brooke, George M., Jr. **John M. Brooke: Naval
      Scientist and Educator.** Charlottesville:
      University Press of Virginia, 1980.

308. Brown, Sandborn C. **Benjamin Thompson, Count
      Rumford.** Cambridge, Mass.: M.I.T. Press,
      1979.

      Discusses the life and career of Count
      Rumford. Includes a discussion of the relation-
      ship between his scientific work on heat and his
      inventions.

309. Brownlie, A.D., and M.F. Lloyd Prichard.
      "Professor Fleeming Jenkin, 1833-85." **Oxford
      Economic Papers** 15 (1963): 204-16.

      Provides a summary of the life of one of the
      early professors of engineering in Great Britain.

310. Brunel, Isambard. **The Life of Isambard Kingdom
      Brunel, Civil Engineer.** Introduction by
      L.T.C. Rolt. Newton Abbot, Devon: David &
      Charles, 1971.

      Reprint of the 1870 edition with a new intro-
      duction. Provides the details of the life of one
      of the leading civil engineers in 19th century
      Britain. Gives examples of his use of science in
      his engineering.

311. Bucciarelli, Louis L., and Nancy Dworsky. **Sophie
      Germain: An Essay in the History of the The-
      ory of Elasticity.** Dordrecht: D. Reidel,
      1980.

      Discusses the work of a significant contribu-
      tor to the theory of elasticity.

312. Cartan, E. **Gaspard Monge, sa vie, son oeuvre.**
      Paris, 1948.

      Provides a study of the life and work on one
      of the founders of the École Polytechnique.

313. Chaldecott, John A. "Josiah Wedgwood (1730-95)--
      Scientist." **British Journal for the History
      of Science** 8 (1975): 1-16.

      Discusses the scientific work of the great
      industrialist Josiah Wedgwood.

314. Channell, David F. **William John Macquorn
      Rankine.** Edinburgh: Scotland's Cultural
      Heritage, 1986.

      Brief biographical study of Rankine and his
      contributions to applied mechanics, thermodynam-
      ics and naval architecture. Also includes a dis-
      cussion of Rankine's role as Regius Professor of
      Civil Engineering and Mechanics at Glasgow Uni-
      versity during the second half of the nineteenth
      century.

315.   Church, William Conant. **The Life of John
       Ericsson.** 2 volumes. New York: Charles
       Scribner's Sons, 1906.

       Provides a biography of Ericsson and de-
       scribes his contributions to naval architecture,
       the design of the <u>Monitor</u>, the development of the
       screw propeller, and the invention of an air en-
       gine.

316.   Cimblers, Borisas. "Reflections on the Motive
       Powers of a Mind." **Physis** 9 (1967): 393-
       420.

       Provides a biographical study of Sadi Carnot
       and his contributions to thermodynamics.

317.   Cioranescu, Alexandre. **Agustin de Betancourt:
       Su obra técnica y científica.** Consejo Supe-
       rior de Investigaciones Científicas, Insti-
       tuto de Estudios Canarios en la Universidad
       de La Laguna, Monografías, vol. 20, sec. 1:
       Ciencias históricas y geográficas, no. 11.
       La Laguna de Tenerife: Universidad de La
       Laguna, 1965.

       Gives the details of the life and work of an
       important contributor to the classification of
       mechanisms.

318.   Cooley, Mortimer E. **Scientific Blacksmith: The
       Autobiography of Mortimer E. Cooley.** New
       York: American Society of Mechanical Engi-
       neers, 1947.

       The autobiography of a leading engineering
       educator and president of the American Society of
       Mechanical Engineers.

319.   Davenport, William Wyatt. **Gyro! The Life and
       Times of Lawrence Sperry.** New York:
       Scribner, 1978.

       Provides a somewhat popular biography of a
       major contributor to control theory.

320. Davidson, Mark. **Uncommon Sense: The Life and Thought of Ludwig von Bertalanffy (1901–1972), Father of General Systems Theory.** Foreword by R. Buckminster Fuller. Introduction by Kenneth E. Boulding. Los Angeles: Tarcher, 1983.

Biography of a leading figure in general systems theory.

321. Dickinson, Henry W. **James Watt, Craftsman and Engineer.** New York: Kelley, 1967.

Reprint of a 1936 edition. Provides the biographical details of the life and work of James Watt.

322. Duckham, B.F. "John Smeaton, the Father of English Civil Engineering." **History Today** 15 (1965): 200–206.

Provides a brief biography of Smeaton and his contributions to the professionalization of civil engineering.

323. Dupuy, R. Ernest. **Sylvanus Thayer: Father of Technology in the United States.** West Point: U.S. Military Academy, 1958.

Gives the biographical details of Thayer's life. Discusses his role in reorganizing West Point's curriculum on the model of French engineering schools.

324. Durand, William F. **Adventures in the Navy, in Education, Science, Engineering, and in War.** New York: American Society of Mechanical Engineers, 1953.

The autobiography of a leading engineering educator and president of the American Society of Mechanical Engineers.

325. _____. **Robert Henry Thurston.** New York: American Society of Mechanical Engineers, 1929.

Presents the biographical details of a leading American engineering scientist and educator who made a significant contribution to a scientific approach to engineering. Mostly descriptive but the only biography available.

326. Ebeling, Heinrich. **Ferdinand Redtenbacher, Leben und Werk.** Karlsruhe, 1943.

     A study of the life and work of a leading engineering scientist and director of the Karlsruhe Polytechnic.

327. Emmerson, George S. **John Scott Russell: A Great Victorian Engineer and Naval Architect.** London: John Murray, 1977.

     Presents a sympathetic portrait of Scott Russell's life. Analyzes his wave line theory for determining the shape of ship hulls. Describes his contributions to the founding of the Institution of Naval Architects and his role in technical education. Does not completely disprove the contemporary criticism of Russell's career.

328. Fairbairn, William. **The Life of Sir William Fairbairn, Bart. (1877).** Partly written by himself. Edited and completed by William Pole. Newton Abbot, Devon: David & Charles, 1970.

     Reprint of the 1877 edition with a new introduction by A.E. Musson. Provides details of the life of one of the founders of materials testing and a pioneer in an experimental approach to the theory of structures.

329. Farrar, W.V. "Andrew Ure, F.R.S., and the Philosophy of Manufactures." **Notes and Records of the Royal Society of London** 27 (1973): 299-324.

     Discusses the life of Andrew Ure and his ideas of the role of science in manufacturing.

330. Fernandex Ordnonez, Jose A. **Eugene Freyssinet.** Barcelona: Editorial Xarait, 1978.

     Biography of a pioneer in the use of reinforced concrete and the inventor of prestressed concrete. Describes his life and career.

331. Finch, James Kip. "Girard: Strength of Materials." **Consulting Engineer** 22 (August 1963): 114-119.

Presents details of the life of Pierre-Simon
Girard, who wrote one of the first books on the
strength of materials in 1798.

332.   Föppl, August. **Lebenserinnerungen**. Munich and
       Berlin: R. Oldenbourg, 1925.

Autobiography of an important engineering
scientist who held several high academic positions in Germany, contributed to the theory of
elasticity, and was the teacher of Prandtl and
Timoshenko.

333.   Gillispie, Charles C. **Lazare Carnot, Savant**.
       Princeton: Princeton University Press, 1971.

Provides a biographical study of Carnot's
contribution to mechanics and mathematics. Discusses the intellectual connections between
Carnot and his son Sadi Carnot. Argues that
Lazare Carnot's most important work was to gener-
alize the concept of vis viva and to apply it to
a theory of machines. Shows how Carnot's use of
vis viva led him to the conclusion that all
shocks and impacts must be avoided in working machines. Provides photographic reproductions of
three early memoirs on mechanics written by
Carnot.

334.   Gillispie, E.S. "Osborne Reynolds." **Physics Ed-
       ucation** 7 (1972): 427-28.

Discusses the life and work of Osborne
Reynolds and his contributions to fluid dynamics.

335.   Gillmor, C. Stewart. **Coulomb and the Evolution
       of Physics and Engineering in Eighteenth-Cen-
       tury France**. Princeton: Princeton University
       Press, 1971.

Reconstructs Coulomb's career as an engineer
and a scientist. Analyzes how his investigations
of magnetic needles led him to research on torsion and fluid resistance. Argues that his background and education in engineering affected his
approach to scientific and technological problems. Discusses how Coulomb contributed to a new
tradition in engineering by combining experimentation and analytic mechanics. Concludes that

his election to the Académie des Sciences changed
his interests from engineering to physics.

336. Grattan-Guinness, I. **Joseph Fourier, 1768-1830.**
      Cambridge, Mass.: M.I.T. Press, 1972.

      Provides a printed version of Fourier's 1807
      memoir on the diffusion of heat. Includes a com-
      mentary on the work and on Fourier's life as well
      as a bibliography of published material by, and
      relating to Fourier. Focuses on an internal
      analysis of the problem of heat diffusion. Pro-
      vides little specific analysis of the use of
      Fourier's work in engineering.

337. Grosser, Morton. **Diesel, the Man and the Engine.**
      New York: Atheneum, 1978.

      A somewhat popular biography of Rudolf Diesel
      and his invention of the Diesel engine.

338. Hall, R. Cargill. "Theodore Von Karman, 1881-
      1963." **Aerospace Historian** 28 (1981): 253-
      58.

      Presents a brief biography of a pioneer in
      the development of a theory of aerodynamics and
      fluid mechanics.

339. Halle, Gerhard. **Otto Lilienthal, Der erste
      Flieger.** 2nd edition. Dusseldorf: VDI
      Verlag, 1956.

      Provides a biography of Lilienthal and his
      work on the theory of flight. Discusses his ex-
      periments on wind and airflow. Also discusses
      the limitations of his theories.

340. Hankins, Thomas L. **Jean d'Alembert: Science of
      the Enlightenment.** Oxford: Clarendon Press,
      1970.

      A biographical study of d'Alembert that also
      discusses the role of science in the Enlighten-
      ment.

341. Harrison, James. "Bennet Woodcroft at the Royal
      Society of Arts, 1845-1857." **Journal of the
      Royal Society of Arts** 128 (1980): 231-33,
      295-97, 375-79.

Discusses Woodcroft's contribution to the
Royal Society of Arts. He also made a significant
contribution to the theory of machines and engineering education.

342. Haupt, Herman. **Reminiscences of General Herman
      Haupt, Giving Hitherto Unpublished Official
      Orders, Personal Narratives of Important Mil-
      itary Operations, and Interviews with Presi-
      dent Lincoln, Secretary Stanton, General-in-
      Chief Halleck, and with Generals McDowell,
      McClellan, Meade, Hancock, Burnside, and Oth-
      ers in Command of the Armies in the Field,
      and His Impressions of These Men.** Milwaukee:
      Wright & Joys, 1901.

      Details the life of Herman Haupt who was in
charge of railway construction during the Civil
War and who also contributed to the theory of
bridge design.

343. Hays, J.N. "The Rise and Fall of Dionysius
      Lardner." **Annals of Science** 38 (1981): 527-
      42.

      Presents an account of Lardner's life includ-
ing his promotion of scientific education and the
application of science to practical problems.
Discusses his decline as a result of his involve-
ment in a scandal and his pessimism concerning
technological progress.

344. Heyman, Jacques. "Couplet's Engineering Memoirs,
      1726-33." **History of Technology** 1 (1976):
      21-44.

      Describes the contributions to engineering
science published by Couplet in the Memoirs of
the French Academy of Sciences. Includes some
details of his life.

345. Hughes, Thomas Parke. **Elmer Sperry: Inventor
      and Engineer.** Baltimore: Johns Hopkins Uni-
      versity Press, 1971.

      Provides a significant and detailed study of
the career of Elmer Sperry, one of the leading
figures in control engineering and founder of the
company that became Sperry Rand Corporation.
Traces the change from the period of heroic lone
inventors to organized group research. Provides

a study of the process of invention in the 20th-century.

346.   _____, ed. **Lives of the Engineers: Selections from Samuel Smiles.** Cambridge, Mass.: M.I.T. Press, 1966.

Selections from Smiles' 19th-century classic biographies of engineers. Includes biographies of John Rennie and Thomas Telford among others.

347.   Keator, F.W. "Benoit Fourneyron (1802-67)." **Mechanical Engineering** 61 (1939): 295-301.

Discusses the life of Fourneyron and his contributions to waterwheel and turbine design.

348.   Killian, James R., Jr. **The Education of a College President: A Memoir.** Cambridge, Mass.: M.I.T. Press, 1985.

Discusses the experiences of James Killian as president of M.I.T. from 1948 to 1959. Gives an account of his role in developing the Servomechanisms Laboratory. Focuses on his view of technological education.

349.   Korting, Johannes. "Ferdinand Redtenbacher." **Zeitschrift des Verein deutscher Ingenieure** 105 (1963): 449-51.

Provides a brief biography of a leading engineering scientist.

350.   Kwik, Robert Julius. "The Function of Applied Science and the Mechanical Laboratory during the Period of Formation of the Profession of Mechanical Engineering, as Exemplified in the Career of Robert Henry Thurston, 1839-1903." Ph.D. dissertation. University of Pennsylvania, 1974.

Traces the career of Thurston. Discusses his role as professor of mechanical engineering at Stevens Institute of Technology and the Engineering School at Cornell University. Also describes his role in establishing a mechanical engineering laboratory.

351.   Lamontagne, Roland. **La Vie et l'oeuvre de Pierre
        Bouguer.** Montreal, 1962.

        Discusses the life and work of a pioneer in
        18th-century naval architecture.

352.   **Leonard Euler, 1707-1783. Beitrage zu Leben und
        Werk.** Basel: Birkhauser Verlag, 1983.

        Contains twenty-seven essays on the life and
        work of Euler. Includes some discussion of his
        work on ships, ballistics and architecture.

353.   Leslie, Stuart W. **Boss Kettering: Wizard of
        General Motors.** New York: Columbia Univer-
        sity Press, 1983.

        Provides a biographical study of an important
        American industrialist who had a great interest
        in the application of science to industry. Dis-
        cusses the role of science in modern industrial
        research.

354.   Leuba, Clarence J. **A Road to Creativity: Arthur
        E. Morgan, Engineer, Educator, Administrator.**
        North Quincy, Mass.: Christopher Publishing
        House, 1971.

        Provides a weak and incomplete biography of
        an important contributor to water projects, in-
        cluding the Miami Conservancy District and the
        TVA, and president of Antioch College.

355.   Lindner, Karl. **Ferdinand Jakob Redtenbacher, der
        Begrunder des wissenschaftlichen Maschinen-
        baues.** Graz, 1959.

        Provides a study of the life of a leading en-
        gineering scientist and his contributions to the
        theory of machines.

356.   Loyrette, Henri. **Gustave Eiffel.** New York:
        Rizzoli, 1985.

        Presents a biographical study of Eiffel as an
        engineer.

357. Nahun, Andrew. **James Watt and the Power of
      Steam**. Hove: Wayland, 1981.

      Discusses James Watt's role in developing the
      steam engine.

358. Nitske, W. Robert, and Charles Morrow Wilson.
      Rudolf Diesel: Pioneer of the Age of Power.
      Norman: University of Oklahoma Press, 1965.

      Provides an account of the life and work of
      Diesel. Discusses the role of theory in the de-
      velopment of his engine. Concludes that Diesel
      was guided to his engine by purely theoretical
      considerations. Somewhat weak in its explanation
      of Diesel's theory.

359. Plank, R. "Franz Grashof als Lehrer und
      Forscher." **Zeitschrift des Vereins deutscher
      Ingenieure** 70 (1926): 28.

      A biography of one of the founders of the VDI
      and the person who succeeded Redtenbacher at
      Karlsruhe.

360. Pritchard, J. Laurence. **Sir George Cayley: The
      Inventor of the Aeroplane**. New York: Horizon
      Press, 1962.

      Presents a biography of Cayley and discusses
      his theory of flight. Includes an analysis of
      his work on the stability of flight and lift-drag
      ratios of wings. Also discusses his role in the
      York Mechanics' Institute and the Polytechnic In-
      stitution of London.

361.   , and E.T. Jones, and Clark B. Millikan.
      ──"Theodore Von Karman, Honorary Fellow, 1881-
      1963." **Journal of the Royal Aeronautical So-
      ciety** 67 (October 1963): 611-617.

      Provides details of the life of a significant
      aeronautical engineer and a contributor to the
      theory of fluid dynamics.

362. Pugsley, Sir Alfred, ed. **The Works of Isambard
      Kingdom Brunel**. London: Institution of Civil
      Engineers, 1976.

      Contains nine essays on Brunel. Presents his
      major engineering works, including the Clifton

Suspension Bridge, the Royal Albert Bridge, his
railway work and the Great Eastern steamship.
Discusses Brunel's theoretical work in technology, especially his work in naval architecture.
Argues that Brunel was more of a synthesizer than
an inventor of new technology.

363.   Raman, V.V. "William John Macquorn Rankine,
          1820-1872." **Journal of Chemical Education**
          50 (1973): 274-76.

       Provides a brief biography of a leading engi-
neering scientist and professor of civil engineering and mechanics at the University of
Glasgow.

364.   Reingold, Nathan. "Alexander Dallas Bache: Sci-
          ence and Technology in the American Idiom."
          Technology and Culture 11 (1970): 163-77.

       Provides an analysis of the life of Bache and
his role in the Coast Survey. Discusses the relationship between theory and practice in his
work. Argues that Bache did not see a chasm sep-
arating theory from practice. Shows that he believed routine technical work had to have a research component. Concludes that Bache's work
transcended the classification of pure or applied
science.

365.   Reinhard, Marcel. **Le grand Carnot.** 2 volumes.
          Paris, 1950-52.

       Provides a political biography of Lazare
Carnot, who made significant contributions to the
theory of machines and was the father of Sadi
Carnot.

366.   Rennie, Sir John. **Autobiography of Sir John
          Rennie, F.R.S., Past President of the Insti-
          tution of Civil Engineers.** London: E. & F.N.
          Spon, 1875.

       Details the life and career of John Rennie
and his contributions to a theoretical approach
to civil engineering.

367.   Robinson, Eric H. "James Watt, Engineer and Man
          of Science." **Notes and Records of the Royal
          Society of London** 24, no. 2 (1970): 221-32.

Discusses the scientific and technological
interests of James Watt.

368.   Rolt, L.T.C.   **James Watt.**   London: Batsford,
        1962.

Discusses the events of Watt's life and the
role of science in his invention of the separate
condenser.

369.   Rosman, Holgar.   **Christopher Polhem.**   Translated
        by William A. Johnson.   Hartford, 1963.

Biographical study of an important contributor to the theory and classification of mechanisms and creator of the "mechanical alphabet."

370.   Sharlin, Harold Issadore, and Tiby Sharlin.   **Lord
        Kelvin, the Dynamic Victorian.**   University
        Park, Pa.: Pennsylvania State University
        Press, 1979.

Includes a discussion of Kelvin's contribution to fluid flow, potential theory and energy.
Emphasizes his scientific contributions.   Does
not adequately explain how Kelvin contributed to
the reformulation of physics that was used by
many engineers.

371.   Sittauer, Hans L.   **Nicolaus August Otto und
        Rudolf Diesel.**   Leipzig: Teubner, 1978.

Discusses the lives of Otto and Diesel and
their contributions to the internal combustion
engine.

372.   Skempton, A.W., ed.   **John Smeaton, F.R.S.**
        London: Telford, 1981.

Presents a series of essays on the life and
work of John Smeaton.   Discusses his contributions to the theory of the steam engine.

373.   Sloan, Edward William.   **Benjamin Franklin
        Isherwood, Naval Engineer: The Years as En-
        gineer in Chief, 1861-1869.**   Annapolis: U.S.
        Naval Institute, 1965.

Presents a study of an important contributor
to the application of scientific theory to steam
navigation.   Includes a discussion of his work on

testing steam engines and his role in the design
of screw-driven cruisers.

374. Smiles, Samuel. **Lives of the Engineers, with an
      Account of Their Principal Works; Comprising
      Also a History of Inland Communication in
      Britain.** 3 Volumes. Newton Abbot, Devon:
      David & Charles, 1968.

      Reprint of the 1862 edition with a new intro-
      duction by L.T.C. Rolt. Provides biographical
      details of many 18th- and 19th-century British
      engineers including Watt, Rennie and Telford.

375. Sokolow, Jayme A. "Count Rumford and Late En-
      lightenment Science, Technology, and Reform."
      Eighteenth Century 21 (1980): 67-86.

      Discusses the contributions of Count Rumford
      to science and technology.

376. Smith, Denis. "David Kirkaldy (1820-1897) and
      Engineering Materials Testing." **Transactions
      of the Newcomen Society** 52 (1980-81): 49-65.

      Discusses the work of an early pioneer in the
      area of materials testing.

377. Sutherland, Hugh B. **Rankine: His Life and Times.
      Lecture Delivered before the British Geotech-
      nical Society at the University of Glasgow on
      13 December 1972 to Mark the Centenary of the
      Death of William John Macquorn Rankine.**
      London: Institution of Civil Engineers, 1973.

      Surveys the life and work of one of the
      founders of engineering science and one of the
      earliest professors of engineering in Great
      Britain. Emphasizes his contributions to the
      theory of earth works.

378. Svenska Technnlogforeningen, Stockholm.
      **Christopher Polhem, the Father of Swedish
      Technology.** Translated by William A.
      Johnson. Hartford: Trustees of Trinity
      College, 1963.

      Provides a biography of a significant Swedish
      engineer who made significant contributions to
      the theory of mechanisms through his mechanical
      alphabet. Contains a bibliography of works on

Polhem, mostly in Swedish. Originally published
in Swedish in 1911.

379. Taton, René. **L'Oeuvre scientific de Gaspard
      Monge.** Paris, 1951.

      Describes the life and works of one of the
      founders of the École Polytechnique.

380. Thomas, Donald E., Jr. **Diesel: Technology and
      Society in Industrial Germany.** Tuscaloosa,
      Alabama: The University of Alabama Press,
      1987.

      Discusses the development of the Diesel en-
      gine. Argues that Diesel's philosophy of tech-
      nology and the rise of the engineering profession
      in 19th century Germany influenced his invention.

381. Tribout, H. **Un grand savant: Le général Jean-
      Victor Poncelet.** Paris, 1936.

      Provides a biography of the professor of me-
      chanics applied to machines at the École
      d'Application de l'Artillerie et du Génie at
      Metz.

382. Turner, Trevor. "John Smeaton, R.F.S. (1724-
      1792)." **Endeavour** 33 (January 1974): 29-33.

      Contains biographical details of the life of
      John Smeaton. Includes his interests in science
      and technology.

383. Von Karman, Theodore, with Lee Edson. **The Wind
      and Beyond: Theodore von Karman, Pioneer in
      Aviation and Pathfinder in Space.** Boston:
      Little, Brown & Co., 1967.

      Presents a narrative autobiography of the
      life and work of Theodore Von Karman. Discusses
      his belief that engineering should be based on
      science. Focuses on his efforts to create a sci-
      entific basis for aeronautical engineering.
      Leaves unanswered several important questions
      concerning his life.

384. Ward, James A. **That Man Haupt: A Biography of
      Herman Haupt.** Baton Rouge: Louisiana State
      University Press, 1973.

        Provides a biography of one of the first
Americans to develop a mathematical theory of
bridges.

385. White, Ruth. **Yankee from Sweden: The Dream and
      the Reality in the Days of John Ericsson.**
      New York: Henry Holt & Co., 1960.

        Surveys the life of John Ericsson and his
role in designing the <u>Monitor</u>. Discusses his
work on air engines, steam powered boats and
naval architecture. Briefly discusses his interest in mathematical theory.

See also: 151, 214, 426.

COLLECTED WORKS, PAPERS AND DRAWINGS

386. Ampère, André Marie. **Journal et correspondance
      d'André-Marie Ampère.** Paris: J. Claye,
      1869.

        Papers and letters of an important member of
the École Polytechnique.

387. Cauchy, Augustin Louis. **Oeuvres complètes; pub-
      liées sous la direction scientifique de
      l'Académie des sciences et sous les auspices
      de M. le Ministre de l'instruction publique.**
      27 Volumes. Paris: Gauthier-Villars, 1882-
      1958.

        Contains important papers on the theory of
structures, the theory of elasticity and the
strength of materials from a professor at the
École Polytechnique.

388. Coulomb, Charles Augustin de. **Mémoires.** Paris:
      Gauthier-Villars, 1884.

        Includes works on 18th-century engineering
science, theory of elasticity, hydraulics and the
theory of machines.

389. Euler, Leonhard. **Opera omnia.** Edited by Jacob
      Ackeret. Lusanne, 1911-57.

        Two series of papers containing theoretical
analyses of the effects of water as a source of

power. The first series establishes that the reaction wheel is more efficient than the traditional value for water wheels and casts doubt on
the universal applicability of Parent's analysis.
The second series focuses on Euler's work in engineering and mechanics.

390. Froude, William. **The Papers of William Froude,**
      M.A., LL.D., F.R.S., 1810-1879. With a mem-
      oir by Sir Westcott Abell and an evaluation
      of William Froude's work by R.W.L. Gawn.
      London: Institution of Naval Architects,
      1955.

      Provides the papers of a key figure in the
      development of naval architecture who did impor-
      tant work on the application of the theory of
      waves to ship design and contributed to the the-
      ory of the stability of ships.

391. Huygens, Christiaan. **Ouevres complètes de**
      Christiaan Huygens. The Hague, 1937.

      Presents experimental studies on hydraulics
      and on efforts to apply mechanics to the water-
      wheel.

392. Jenkin, Henry Charles Fleeming. **Papers, Liter-**
      ary, Scientific, &c. Edited by Sidney Colun
      and J.A. Ewing. With a memoir by Robert
      Louis Stevenson. 2 Volumes. London:
      Longmans, Green & Co., 1887.

      Volume 1 includes a biographical memoir of
      Jenkin and his role as professor of engineering
      at the University of Edinburgh. Volume 2 con-
      tains papers on technical education and on the
      application of graphical methods to the determi-
      nation of the efficiency of machinery.

393. Kelvin, William Thomson, Baron. **Mathematical and**
      Physical Papers. 6 Volumes. Cambridge:
      Cambridge University Press, 1882-1911.

      Includes papers on elasticity, thermodynamics
      and fluid dynamics.

394. Lagrange, Joseph Louis. **Oeuvres de Lagrange.** 14
      Volumes. Paris: Gauthier-Villars, 1867-92.

Collected works of a leading 18th-century
scientist whose work influenced applied mechanics
and engineering. Includes works on analytical
mechanics. Lagrange taught at the École Polytechnique.

395. Lilienthal, Otto. **Papers on Aeronautics with
      Some Translation.** Edited by Octave Chanute.
      Chicago: John Crerar Library, 1952.

      Collected papers of one of the pioneers in
the study of aeronautics.

396. McFarland, Marvin W., ed. **The Papers of Wilbur
      and Orville Wright.** 2 volumes. New York:
      McGraw-Hill, 1953.

      Volume 1 covers the period 1899 to 1905 and
volume 2 the period 1906 to 1948. Includes work
on the theory of flight.

397. Maxwell, James Clerk. **The Scientific Papers of
      James Clerk Maxwell.** 2 Volumes. Edited by
      W.D. Niven. Cambridge: Cambridge University
      Press, 1890.

      Includes papers on the kinetic theory of
gases and on the theory of matter, both of which
influenced engineering thermodynamics.

398. Molella, Arthur P., and Nathan Reingold, Marc
      Rothenberg, Joan F. Steiner, Kathleen
      Waldenfels. **A Scientist in American Life:
      The Essays and Lectures of Joseph Henry.**
      Foreword by Lewis Thomas. Washington, D.C.:
      Smithsonian Institution Press, 1980.

      Contains a collection of essays and lectures
by Joseph Henry. Argues for the role of theoretical science as the basis for technological
progress.

399. Parent, Antoine. **Essais et recherches de mathé-
      matique et de physique.** 3 Volumes. Paris:
      J. de Nully, 1713.

      Collected works of Parent. Includes early
scientific analysis of machines and water wheels.

400. Parsons, Charles A. **The Scientific Papers and
      Addresses of the Hon. Charles A. Parsons.**
      Edited by G.L. Parsons. Cambridge: Cambridge
      University Press, 1934.

      Contains fundamental work on the development
      of the steam turbine and its application in navi-
      gation.

401. Rankine, William John Macquorn. **Miscellaneous
      Scientific Papers.** With a biographical mem-
      oir by P.G. Tait. Edited by W.J. Millar.
      London: Charles Griffin & Co., 1881.

      Some of the collected papers of one of the
      founders of engineering science and Regius Pro-
      fessor of Civil Engineering and Mechanics at
      Glasgow University. Part I contains papers re-
      lating to temperature, elasticity, and expansion
      of vapors, liquids and solids. Part II contains
      papers relating to energy and its transforma-
      tions, thermodynamics, and the mechanical action
      of heat in the steam engine. Part III contains
      papers relating to wave forms, propulsion of ves-
      sels, and the stability of structures.

402. Reingold, Nathan, ed. **The Papers of Joseph
      Henry.** Volume I: **The Albany Years, December
      1797-October 1832.** Volume II: **The Princeton
      Years, November 1832-December 1835.** Volume
      III: **The Princeton Years, January 1936-Decem-
      ber 1837.** Volume IV: **The Princeton Years,
      January 1838-December 1840.** Volume V: **The
      Princeton Years, January 1841-December 1843.**
      Washington, D.C.: Smithsonian Institution
      Press, 1972-85.

      First of a projected 15-volume series on the
      papers of a leading American scientist and first
      Secretary of the Smithsonian Institution, and a
      contributor to American technology. Material is
      highly selected and annotated. Includes a dis-
      cussion of Henry's study of the application of
      theory to technology. Also includes a discussion
      of steam as an energy source and a discussion of
      the construction of ships. Later editions will
      be edited by Marc Rothenberg.

403. Reynolds, Osborne. **Papers on Mechanical and
      Physical Subjects.** Cambridge: Cambridge Uni-
      versity Press, 1900-1903.

Reprints of articles by a leading engineering scientist. Includes works on the theory of the steam engine, steam engine indicators, and the mechanical equivalent of heat.

404. Robinson, Eric, and Douglas McKie, eds. **Partners in Science: Letters of James Watt and Joseph Black.** Cambridge, Mass.: Harvard University Press, 1970.

Contains the correspondence between Joseph Black and James Watt. Discusses Watt's interest in science. Includes materials on the development of the steam engine.

405. Rumford, Count. **The Collected Works of Count Rumford.** Volume 1: **The Nature of Heat.** Volume 2: **Practical Application of Heat.** Volume 3: **Devices and Techniques.** Volume 4: **Light and Armament.** Volume 5: **Public Institutions.** Edited by Sandborn C. Brown. Cambridge, Mass.: Harvard University Press, 1968-70.

Presents the works of the American émigré Benjamin Thompson, Count Rumford, who contributed new insights about heat and helped establish the Royal Institution in London. Discusses experiments on heat and the application of theories of heat to practical devices such as stoves and fireplaces. Shows how Rumford used scientific discoveries for technological advances.

406. Smeaton, John. **Reports of the Late John Smeaton.** 4 volumes. London, 1812-14.

Contains the engineering reports of one of the leading civil engineers of the 18th century. Includes works on building and hydraulics, and water power.

407. Stapleton, Darwin H., ed. **The Engineering Drawings of Benjamin Henry Latrobe.** New Haven: Yale University Press, 1980.

A collection of Latrobe's engineering drawings with annotations and an introductory essay by the editor. Argues that Latrobe introduced Americans to a new standard of engineering drawing by drawing to scale and by adding realistic shading and details to his drawings. Drawings

are analyzed as examples of "nonverbal thought."
Drawings provide an "engineering vocabulary" of
early 19th-century American engineering.

408. Stokes, George Gabriel. **Mathematical and Physi-
      cal Papers.** 5 Volumes. Cambridge: Cambridge
      University Press, 1880-1905.

      Includes works on hydrodynamics and stream-
      line theory. Influenced the work of naval archi-
      tects.

409. _____. **Memoir and Scientific Correspondence of
      the Late George Gabriel Stokes.** 2 Volumes.
      Edited by Joseph Larmor. Cambridge:
      Cambridge University Press, 1907.

      Contains a biographical study of Stokes and
      discusses his contributions to hydrodynamics.

410. Tann, Jenifer. **The Selected Papers of Boulton
      and Watt.** Volume I: **The Engine Partnership,
      1775-1825.** Cambridge, Mass.: M.I.T. Press,
      1981.

      Presents details of the early history of the
      steam engine. Includes insights into the role of
      science in the development of the steam engine.

411. Thomson, James. **Collected Papers in Physics and
      Engineering.** Edited by Joseph Larmor.
      Cambridge: Cambridge University Press, 1912.

      Collected papers of James Thomson, Professor
      of Engineering at Queen's University, Belfast and
      later at Glasgow University. Includes works on
      fluid dynamics, and the theory of elasticity.
      Brother of Lord Kelvin.

412. Watt, James. **The Origin and Progress of the Me-
      chanical Inventions of James Watt. Illus-
      trated by His Correspondence with His Friends
      and the Specifics of His Patents.** 3 volumes.
      London: J. Murray, 1854.

      Collected letters of James Watt with discus-
      sion of his invention of improvements of the
      steam engine.

CHAPTER IV: INSTITUTIONS

THE HISTORY OF ENGINEERING EDUCATION

413.   Ahlstrom, Goran. **Engineers and Industrial
       Growth: Higher Technical Education and the
       Engineering Profession during the 19th and
       Early 20th Centuries: France, Germany,
       Sweden, and England.** London: Croom Helm,
       1982.

       Studies the relationship between technical
       education and the growth of industry. Compares
       engineering education in England and on the Con-
       tinent in the 19th century.

414.   _____.   "Higher Technical Education and the En-
       gineering Profession in France and Germany
       during the 19th Century: A Study on Techno-
       logical Change and Industrial Performance."
       Economy and History 21 (1978): 51-88.

       Includes a discussion of the relationship be-
       tween technical education and the professional-
       ization of engineering.

415.   Alexander, Daniel E. "The Development of Engi-
       neering Education in the United States."
       Ph.D. dissertation. Washington State Univer-
       sity, 1977.

       Includes a brief survey of engineering educa-
       tion in Europe. Traces European influences on
       American engineering education.

416.   Argles, Michael. **South Kensington to Robbins:
       An Account of English Technical and Scien-
       tific Education since 1851.** London:
       Longmans, 1964.

Presents a useful description and social history of English scientific and technical education. Discusses the interaction of science and technology.

417. Armytage, W.H.G. **Civic Universities: Aspects of a British Tradition.** London: Benn, 1955.

Includes a discussion of the rise of polytechnic institutes in the second half of the 19th century.

418. Arrowsmith, H. **Pioneering in Education for Technologies: The Story of Battersea College of Technology, 1891-1962.** London: University of Surrey, 1966.

Provides a narrative description of Battersea College of Technology.

419. Artz, Frederick B. **The Development of Technical Education in France, 1500-1850.** Cambridge, Mass.: M.I.T. Press for the Society for the History of Technology, 1966.

Presents a useful analysis of the role of technology in French education. Includes a discussion of the role of the École Polytechnique in establishing a framework for engineering science.

420. Ashby, Eric. **Technology and the Academics: An Essay on Universities and the Scientific Revolution.** London: Macmillan & Co., 1959.

Discusses the impact of the scientific and industrial revolutions on British universities. Includes a discussion of the role of science and technology in the university curriculum. Important essay.

421. Aubry, Jacques. "Création des instituts de sciences appliquées par la Faculté des Sciences de Nancy de 1890 à 1919." **Comptes rendus 103 Congrès national des Sociétés Savantes, section des sciences** 5 (1978): 173-80.

Discusses the creation of institutes of applied science by the science faculty at Nancy.

422.   Bailey, Bill. "The Development of Technical Edu-
       cation, 1934-1939." **History of Education** 16
       (1987): 49-65.

423.   Baker, Donald N., and Patrick J. Harrigan, eds.
       **The Making of Frenchmen: Current Directions
       in the History of Education in France, 1679-
       1979.** Waterloo, Ontario: Historical Reflec-
       tions Press, 1980.

       Includes articles on the Écoles d'Arts et
Métiers, and on the role of scientific education.

424.   Barlow, Melvin E. **History of Industrial Educa-
       tion in the United States.** Peoria: Bennett,
       1967.

       Provides a survey of industrial education in
America.

425.   Bellot, H. Hale. **University College London,
       1826-1926.** London: University of London
       Press, 1929.

       Includes a discussion of the attempts to es-
tablish a chair of engineering and the relationship between engineering and natural philosophy.

426.   Berth, Donald F., and Gladys J. McConkey.
       "Capstones of Century One." **Engineering:
       Cornell Quarterly** 6 (Autumn 1971): 1-100.

       Provides a history of the Cornell School of
Engineering including the work of Robert Thurston
during the 19th century.

427.   Birse, Ronald M. **Engineering at Edinburgh Uni-
       versity: A Short History, 1673-1983.**
       Edinburgh: School of Engineering, Edinburgh
       University, 1983.

       Discusses the teaching of engineering at
Edinburgh and the establishment of a professorship of technology during the 19th century.

428.   Bishop, Morris. **A History of Cornell.** Ithaca:
       Cornell University Press, 1962.

       Includes a discussion of the history of the
Silbey College of Engineering at Cornell and the
role of Robert Thurston in its development.

429. Bradley, Margaret. "Civil Engineering and Social
      Change: The Early History of the Paris École
      des ponts et chaussées." **History of Educa-
      tion** 14 (1985): 171-83.

      Provides the early history of an important
      French engineering school.

430. _____. "The Facilities for Practical Instruc-
      tion in Science during the Early Years of the
      École Polytechnique." **Annals of Science** 33
      (1976): 425-46.

      Discusses how science was taught at the École
      Polytechnique.

431. _____. "Scientific Education for a New Society:
      The École Polytechnique, 1795-1830." **History
      of Education** 5 (1976): 11-24.

      Discusses the relationship between the École
      Polytechnique and French society.

432. Brittain, James E., and Robert C. McMath, Jr.
      "Engineers and the New South Creed: The For-
      mation and Early Development of Georgia
      Tech." **Technology and Culture** 18 (1977):
      175-201.

      Argues that engineering education in the
      South reflected a desire for recovery, moderniza-
      tion, and restoration of Southern power. Ana-
      lyzes the development of Georgia Tech in terms of
      Monte Calvert's distinction between school cul-
      ture and shop culture (see item 533). Concludes
      that Georgia Tech was founded on the shop culture
      tradition which was practical and fit with the
      New South Creed, but by the 1890s it was moving
      toward a school culture tradition.

433. Brunot, A. and R. Coquand. **Le Corps des Ponts et
      Chaussées: Histoire de l'administration
      française.** Paris: Editions du Centre Na-
      tional de la Recherche Scientifique, 1982.

      An official history of the role of the school
      in developing the transportation system of
      France. Contains significant documentation but
      not much critical analysis.

434.   Buchanan, R.A. "Science and Engineering: A Case
       Study in British Experience in the Mid-Nine-
       teenth Century." **Notes and Records of the
       Royal Society of London** 32 (1978): 215-23.

       Describes the types of education that were
       available to I.K. Brunel and other engineers dur-
       ing the mid-19th century.

435.   Cardwell, Donald S.L., ed. **Artisan to Graduate:
       Essays to Commemorate the Foundation in 1824
       of the Manchester Mechanics' Institution, Now
       in 1974 the University of Manchester Insti-
       tute of Science and Technology.** Manchester:
       Manchester University Press, 1974.

       Presents seventeen essays on the development
       of modern technological education. Discusses
       technical education in France, Germany, Scotland
       and England. Details the origins and early years
       of the Manchester Mechanics' Institution.

436.   _____. **The Organisation of Science in England.**
       London: William Heinemann, 1957.

       Discusses the role of science and technology
       in established universities such as Oxford and
       Cambridge, as well as their role in the dissent-
       ing academies and mechanics' institutes. Focuses
       on England with no discussion of Scotland or
       Ireland.

437.   Chittenden, Russell. **History of the Sheffield
       Scientific School of Yale University, 1846-
       1922.** 2 Volumes. New Haven: Yale University
       Press, 1928.

       Provides a history of the Sheffield Scien-
       tific School at Yale. Includes a discussion of
       the role of engineering in the school.

438.   Cotgrove, Stephen F. **Technical Education and So-
       cial Change.** London: George Allen & Unwin,
       1958.

       Discusses the role of mechanics' institutes
       in bringing together science and technology.

439.   Coutts, James. **A History of the University of
       Glasgow: From Its Foundation in 1451 to 1909.**
       Glasgow: James Maclehose & Sons, 1909.

Includes a discussion of the creation of a
chair of engineering and its relationship to the
natural sciences.

440.   Crosland, M.P., ed. **Science in France in the**
       Revolutionary Era, Described by Thomas Bugge.
       Cambridge, Mass.: M.I.T. Press, 1969.

       Describes the state support of science during
       the French Revolution. Includes descriptions of
       the Écoles Centrales, the École Polytechnique,
       military schools, mining schools and schools of
       civil engineering.

441.   Davie, George Elder. **The Democratic Intellect:**
       Scotland and Her Universities in the Nine-
       teenth Century. Edinburgh: University of
       Edinburgh Press, 1961.

       Includes a very profound analysis of the role
       of engineering within the Scottish universities
       and its relationship to science. Discusses how
       science, as shaped by Scottish Common Sense phi-
       losophy, was viewed as practical and useful.

442.   Day, C.R. **Education for the Industrial World:**
       The Écoles d'Arts et Métiers and the Rise of
       French Industrial Engineering. Cambridge,
       Mass.: M.I.T. Press, 1987.

       Provides a social history of the écoles
       d'arts et metiers and their relationship to the
       grandes écoles such as the École Polytechnique.
       Discusses the role of status in defining the
       écoles.

443.   _____. "The Making of Mechanical Engineers in
       France: The Écoles d'arts et métiers, 1803-
       1914." **French Historical Studies** 10 (1978):
       439-60.

       Questions the assumption that French engi-
       neering schools produced mostly theoretically
       trained engineers and few practical engineers.

444.   _____. "Technical and Professional Education in
       France: The Rise and Fall of L'Enseignement
       secondaire special, 1865-1902." **Journal of**
       Social History 6 (1972-73): 177-201.

445. **École Polytechnique: Livre du centenaire, 1794-1894.** 3 volumes. Paris: Gauthier-Villars, 1895-97.

    Massive study of the École Polytechnique including the biographies of many of the faculty.

446. Edmonson, James M. "From mécanicien to ingénieur: Technical Education and the Machine-Building Industry in 19th Century France." Ph.D. dissertation. University of Delaware, 1981.

    Investigates the changing character and source of technical skills and knowledge. Describes the rise of technical education. Concludes that before 1850 the techniques of machine builders were rooted in a craft tradition but by 1880 machine builders acknowledged the value of technical education.

447. Elliot, Arlene Ann. "The Development of the Mechanics' Institutes and Their Influence upon the Field of Engineering: Pennsylvania, a Case Study, 1824-1860." Ph.D. dissertation. University of Southern California, 1972.

448. Emmerson, George S. **Engineering Education: A Social History.** Newton Abbot, Devon: David & Charles, 1973.

    Provides a narrative description of engineering education throughout the world. Emphasizes 19th-century Europe.

449. Finch, James Kip. **A History of the School of Engineering, Columbia University.** New York: Columbia University Press, 1954.

    Narrative history of the School of Engineering at Columbia.

450. Fisher, Berenice M. **Industrial Education: American Ideals and Institutions.** Madison: University of Wisconsin Press, 1967.

451. Forman, Sidney. **West Point: A History of the United States Military Academy.** New York, 1950.

Presents a limited account of one of the
first schools in America to teach engineering
based on scientific principles.

452. Fourcy, Ambroise. **Histoire de l'École Polytech-
      nique.** Introduction by Jean Dhombres.
      Paris: Belin, 1987.

Modern reprint of the 1828 work. One of the
first histories of the École Polytechnique. Focuses on the founding of the school. Dhombres'
introduction provides a review of the historical
studies on the École Polytechnique.

453. Furman, Franklin, ed. **Morton Memorial: A History
      of the Stevens Institute of Technology.**
      Hoboken, N.J.: Stevens Institute of Technol-
      ogy, 1905.

Includes a history of the institute along
with biographies of faculty members.

454. Garner, A.D., and E.W. Jenkins. "The English Me-
      chanics' Institutes: The Case of Leeds,
      1824-42." **History of Education** 13
      (1984):139-52.

Discusses the establishment of a mechanics'
institute at Leeds and its role in providing scientific training to mechanics.

455. Gispen, Corneliss W.R. "Technical Education and
      Social Status: The Emergence of the Mechani-
      cal Engineering Occupation in Germany, 1820-
      1890." Ph.D. dissertation. University of
      California at Berkeley, 1981.

456. Gordon, Dane R. **Rochester Institute of Technol-
      ogy: Industrial Development and Educational
      Innovation in an American City.** New York:
      Edwin Mellen Press, 1982.

Discusses the history of Rochester Institute
of Technology and its relation to industrial development in Rochester, New York.

457. Grattan-Guinness, I. "Work for the Workers: Ad-
      vances in Engineering Mechanics and Instruc-
      tion in France, 1800-1830." **Annals of Sci-
      ence** 41 (1984): 1-33.

Discusses the emergence of the concept of work in mechanics' textbooks during the first part of the 19th century.

458. Grayson, Lawrence P. "A Brief History of Engineering Education in the United States." **Engineering Education** 68 (1977): 246-64.

Presents a survey of engineering education beginning with West Point. Includes a discussion of curricula.

459. Hall, A. Rupert. **Science for Industry: A Short History of the Imperial College of Science and Technology.** London: Imperial College of Science and Technology, 1982.

Commemorative study prepared for the College's 75th anniversary. Discusses how the ambivalent relationship between science and practice in the 19th century affected the transformation of the Royal School of Mines and the Royal College of Chemistry into Imperial College. One of the first integrated histories of Imperial College, but it lacks a detailed study of the Royal School of Mines and the Royal College of Chemistry. Does not contain footnotes, bibliography or index.

460. Heydon, Roy. "The Glasgow Mechanics' Institution." **The Philosophical Journal** (Transactions of the Royal Philosophical Society of Glasgow) 10 (1973): 107-120.

Describes the origins, role, and curriculum at the Glasgow Mechanics' Institution. Discusses the role played by the Institution in teaching science and mathematics to engineers and mechanics. Emphasizes the social history of the Institution.

461. Hilken, T.J.N. **Engineering at Cambridge University, 1783-1965.** Cambridge: Cambridge University Press, 1967.

Provides a useful study of the development of engineering at Cambridge. Discusses the interaction of science and technology through the role of the Jacksonian Professorship, a forerunner of the chair of engineering.

462. Inkster, Ian. "Science and the Mechanics' Insti-
      tutes, 1820-1850: The Case of Sheffield."
      Annals of Science 32 (1975): 451-74.

      Argues that mechanics' institutes were a
      product of a widespread scientific culture.

463. Institution of Civil Engineers. **The Education
      and Status of Civil Engineers in the United
      Kingdom and in Foreign Countries.** London:
      Institution of Civil Engineers, 1870.

      Provides a survey of technical education.
      Includes an outline of scientific and engineering
      courses taught at various universities and the
      requirements for degrees or certificates.

464. Jacob, Dillard. **102 Years: A Story of the First
      Century of Vanderbilt University School of
      Engineering, 1875-1975.** Nashville:
      Vanderbilt Engineering Alumni Association,
      1975.

      Provides a narrative history of the engineer-
      ing school at Vanderbilt.

465. Japiot, Francois. "Les Bourguignons et l'École
      Polytechnique." **Mémoires de l'Académie des
      Sciences, Arts et Belles-Lettres de Dijon**
      123 (1976-78): 277-97.

      Describes the role of the Bourguignons in the
      founding of the École Polytechnique.

466. Kastner, Richard H. "Die Entwicklung von Technik
      und Industrie in Österreich und die Technis-
      che Hochschule in Wien." **Blätter für Tech-
      nikgeschichte** 26 (1965): 1-186.

      Discusses the history of technical education
      in Austria. Focuses on the Vienna Engineering
      College, founded in 1815.

467. Kelly, Thomas. **A History of Adult Education in
      Great Britain.** Liverpool: University of
      Liverpool Press, 1962.

      Discusses the history of the mechanics' in-
      stitutes.

468. Killian, James R. "Centenary of the
     Massachusetts Institute of Technology."
     Nature 190 (1961): 948-53.

     Discusses the founding and the history of
     M.I.T.

469. Knoll, H.B. **The Story of Purdue Engineering.**
     West Lafayette, Ind.: Purdue University
     Press, 1963.

     Traces the history of engineering at Purdue.

470. Langins, Janis. **La République avait besoin de
     savants: Les debuts de l'École Polytechnique:
     L'École centrale des travaux publics et les
     cours révolutionnaires de l'An III.** Preface
     by Emmanuel Grison. Paris: Belin, 1987.

     Focuses on the École Polytechnique's first
     year. Based on the diary of the school's first
     student inspector. Describes the daily life at
     the school. Includes a survey of the <u>cours revo-
     lutionnairs</u>.

471. _____. "The École Polytechnique (1794-1804):
     From Encyclopaedic School to Military Insti-
     tution." Ph.D. dissertation. University of
     Toronto, 1979.

     Analyzes the early years of the École Poly-
     technique.

472. _____. "Sur la première organisation de l'École
     polytechnique: Texte de l'arrête du 6
     frimaire an III." **Revue d'Histoire des Sci-
     ences et de leurs Applications** 33 (1980):
     289-313.

     Discusses the founding of the École Polytech-
     nique.

473. Léon, Antoine. **Histoire de l'education tech-
     nique.** Paris: Presses universitaires de
     France, 1961.

     Contains a brief discussion of the history of
     technical education in France during the past two
     hundred years.

474. _____. **La Révolution Française et l'education technique.** Paris: Société des études robespierristes, 1968.

Discusses the relationship between technical education and the French Revolution.

475. Lloyd, B.E., and W.J. Wilkin. **The Education of Professional Engineers in Australia.** 2nd edition. N.p.: Association of Professional Engineers, Australia, 1962.

Includes a chapter on the history of engineering education in Australia.

476. Lundgreen, P. "Education for the Science-Based Industrial State: The Case for Nineteenth Century Germany." **History of Education** 13 (1984): 59-67.

Discusses the relationship between education and industrial development in Germany.

477. McMath, Robert C., Jr., et al. **Engineering the New South: Georgia Tech, 1885-1985.** Athens: University of Georgia Press, 1985.

Discusses the development of Georgia Tech. and the debate over the role of science in engineering education.

478. Manegold, Karl-Heinz. "Technology Academised: Education and Training of the Engineers in the Nineteenth Century." **The Dynamics of Science and Technology** (item 38), pp. 137-58.

Discusses the institutionalization of engineering in 19th-century German universities. Describes the status disputes between universities and institutes of technology. Analyzes the role of the state and industry in the eventual attempt at the equalization of rank between engineers and scientists.

479. Martin, Claude Trimble. **From School to Institute; an Informal Story of Case.** Cleveland: World Publishing, 1967.

Provides a history of Case Institute of Technology which began as Case School of Applied Science.

480. Molly, Peter Michael. "Technical Education and
      the Young Republic: West Point as America's
      École Polytechnique, 1802-1833." Ph.D. dis-
      sertation. Brown University, 1975.

      Discusses how the École Polytechnique became
      the model for engineering education at West
      Point.

481. Morton, R.G. "Mechanics' Institutes and the At-
      tempted Diffusion of Useful Knowledge in
      Ireland, 1825-79." **Irish Booklore** 2 (1972):
      59-74.

      Discusses the role of mechanics' institutes
      in fostering an interest in the mechanical arts
      in Ireland.

482. Murray, David. **Memories of the Old College of
      Glasgow.** Glasgow: Jackson, Wylie & Co.,
      1927.

      Discusses the establishment of the first
      chair of engineering at a British university.

483. Nicol, John. **The Technical Schools of New
      Zealand: An Historical Survey.** N.p.: New
      Zealand Council for Educational Research,
      1940.

484. Oakley, C.A. **History of a Faculty.** Glasgow:
      University of Glasgow, 1973.

      Describes the history of engineering at the
      University of Glasgow and the establishment of
      the first chair of engineering in Great Britain.
      Includes discussions of W.J.M. Rankine and James
      Thomson.

485. Pippard, A.J.S. "Education and Theory." **The En-
      gineer Centenary Number** (1956): 161-63.

      Useful survey of the rise of engineering edu-
      cation during the 19th century. Focuses on
      Britain. Also discusses the rise of British en-
      gineering societies.

486. Prescott, Samuel C. **When M.I.T. Was "Boston
      Tech" 1861-1916.** Cambridge, Mass.: M.I.T.
      Press, 1954.

Discusses the early years of M.I.T.

487. Rabkin, Yakov M., and J. Ann Levi-Lloyd.
      "Technology and Two Cultures: 100 Years of
      Engineering Education in Montreal." **Minerva**
      22 (1984):67-97.

488. Reinhold, Meyer. "The Quest for 'Useful Knowl-
      edge' in Eighteenth Century America." **Pro-
      ceedings of the American Philosophical Soci-
      ety** 119, no. 2 (1975): 108-132.

      Includes a discussion of the introduction of
the mechanical arts into the curriculum.

489. Rezneck, Samuel. **Education for a Technological
      Society: A Sesquicentennial History of
      Rensselaer Polytechnic Institute.** Troy,
      N.Y.: Rensselaer Polytechnic Institute, 1968.

      Provides a good institutional history of one
of the earliest engineering schools in America.

490. Ricketts, Palmer C. **History of Rensselaer Poly-
      technic Institute, 1824-1914.** 3rd edition.
      New York: John Wiley & Sons, 1934.

      Presents a useful description of the institu-
tionalization of engineering. Provides a list of
courses given at the school.

491. Robertson, Paul L. "Technical Education in the
      British Shipbuilding and Marine Engineering
      Industries, 1863-1914." **Economic History Re-
      view** 27 (1974): 222-35.

492. Roderick, Gordon W., and Michael D. Stephens.
      "Education and Training for English Engineers
      in the Late 19th Century and Early 20th Cen-
      tury." **Annals of Science** 27 (1971): 143-63.

      Focuses on Liverpool as a case study for the
rest of England.

493. _____. "The Late Victorians and Scientific and
      Technical Education." **Annals of Science** 28
      (1972): 385-400.

      Argues that Britain's loss of economic lead-
ership in the late 19th century resulted from the

failure to provide the right kind of industrial education.

494.   _____. "Science and Technology at English Universities and Colleges and the Economic Development during the 19th Century." **Technikgeschichte** 40, no. 3 (1973): 226-50.

495.   _____. **Scientific and Technical Education in Nineteenth Century England: A Symposium.** Newton Abbot, Devon: David & Charles, 1973.

Contains nine reprinted articles.

496.   _____. "Science, the Working Classes and Mechanics' Institutes." **Annals of Science** 29 (1972): 349-60.

Discusses the role of mechanics' institutes in fostering interest in science among technologists.

497.   Rürup, Reinhard, ed. **Wissenshaft und Gesellschaft Beiträge zur Geschichte der Technischen Universität Berlin, 1879-1979.** Berlin: Springer, 1979.

Collection of essays on the Technical University of Berlin.

498.   Sanderson, Michael. **The Universities and British Industry, 1850-1970.** London: Routledge & Kegan Paul, 1972.

Discusses the role of the universities and of science in bringing about modern engineering.

499.   Shapin, Steven, and Barry Barnes. "Science, Nature and Control: Interpreting Mechanics' Institutes." **Social Studies of Science** 7 (1977): 31-74.

Discusses the role of science in the mechanics' institutes. Argues that the founders thought science would aid in the social control of artisans.

500.   Shinn, Terry. **Savoir scientifique et pouvoir social: L'École polytechnique, 1794-1914.** Paris: Presses de la Fondation National des Sciences Politiques, 1980.

Discusses the history of the École Polytechnique during the 19th century. Provides a quantitative analysis of the graduates during the
19th century. Shows how Laplace's deductivist
approach to the instruction of students became
dominant over Monge's more practical approach.
Concludes that the emphasis on esoteric knowledge
over experimental techniques led the new chemical
and electrical industries to prefer graduates of
the École Centrale over those of the École Polytechnique.

501.   _____. "Des sciences industrielles aux sciences
         fondamentales: La mutation de l'École su-
         périeure de physique et de chimie (1882-
         1970)." **Revue française de sociologie** 22
         (1981): 167-82.

Discusses a debate between supporters of pure
science and supporters of applied science at the
École Supérieure de Physique et de Chimie.

502.   Simpson, Renate. **How the Ph.D. Came to Britain:
         A Century of Struggle for Postgraduate Educa-
         tion.** Guildford, Surrey: Society for Re-
         search into Higher Education, 1983.

Argues that the needs of industry were important in the development of postgraduate education.

503.   Small, James. "Engineering." **Fortuna Domus: A
         Series of Lectures Delivered in the Univer-
         sity of Glasgow in Commemoration of the Fifth
         Centenary of Its Foundation.** Glasgow: The
         University of Glasgow Press, 1952, pp. 335-
         355.

Provides a useful discussion of the establishment of the first chair of engineering in
Great Britain and the relationship between the
professor of engineering and the professors of
natural science and mathematics.

504.   Stephens, Michael D. "British Artisan, Scien-
         tific and Technical Education in the Early
         19th Century." **Annals of Science** 29 (1972):
         87-98.

Discusses the role of the mechanics' institutes in providing scientific and mathematical
training for technologists.

505. _____. "Changing Attitudes to Education in
England and Wales, 1833-1902: The Government
Reports, with Particular Reference to Science
and Technical Studies." **Annals of Science**
30 (1973): 149-64.

506. Sterland, E.G. "The Early History of the Teaching of Engineering in Cambridge." **Transactions of the Newcomen Society** 28 (1951-53):
263-75.

Discusses the interaction of science and
technology through the Jacksonian Professorship.

507. Stine, Jeffrey K. "Professionalism vs. Special
Interest: The Debate Over Engineering Education in 19th Century America." **Potomac Review** 26-27 (1984-85): 72-94.

508. Strunz, Hugo. **Von der Bergakademie zur Technischen Universität Berlin, 1770-1970.** Essen:
Glückauf, 1970.

Provides a history of the Technical University in Berlin.

509. Stutzner, Heinz, and Dagmar Szollosi. "The Development of Technical Education during the
Second Stage of the Industrial Revolution in
Saxony." **History and Technology** 2 (1985):
269-82.

510. Sullivan, F.B. "The Royal Academy at
Portsmouth." **Mariners Mirror** 63 (November
1977): 311-26.

Discusses a forerunner of the Royal Naval
College. Provides details of the study of navigation, gunnery and fortification based on applied mathematics and science.

511. Taton, René, ed. **Enseignement et Diffusion des
Sciences en France au XVIIIe siècle.** Paris:
Hermann, 1964.

Includes a discussion of the École Royale des
Ponts et Chaussées, the École des Mines, and the

engineering school at Mézières. Also includes
discussions of ship construction, military engineering, design.

512. Timmons, George. "Education and Technology in
the Industrial Revolution." **History of Technology** 8 (1983): 135-49.

Discusses the relationship between technology
and education during the period of the Industrial
Revolution.

513. _____. "Science, Technology, and Education in
the Industrial Revolution." **Endeavour** 10
(1986): 85-89.

Argues against the idea that the leaders of
the Industrial Revolution had little education.
Concludes that a variety of educational institutions were influential at the time.

514. Timoschenko, Stephen P. "The Development of Engineering Education in Russia." **The Russian
Review** 15 (1956): 173-85.

Discusses how Russia established the equivalent of the École des Ponts et Chaussées.

515. _____. **Engineering Education in Russia**. New
York: McGraw-Hill, 1959.

One of the few studies in English on engineering education in Russia.

516. Tylecote, Mabel. **The Mechanics' Institutes of
Lancashire and Yorkshire before 1851**.
Manchester: Manchester University Press,
1957.

Discusses the scientific and mathematical
training provided to mechanics by the Institutes.

517. Wantabe, Minoru. "Japanese Students Abroad and
the Acquisition of Scientific and Technical
Knowledge." **Cahiers d'histoire mondiale** 9,
no. 2 (1965) 254-293.

Focuses on the second half of the 19th century.

518. Weiss, John Hubbel. **The Making of Technological
Man: The Social Origins of French Engineer-**

ing Education. Cambridge, Mass.: M.I.T.
Press, 1982.

Focuses on the early history of the École
Centrale des Arts et Manufactory (founded 1829).
Details the French attitudes toward pure and ap-
plied science. Makes some comparisons to the
École Polytechnique. Discusses the social back-
ground of engineering students.

519. Weisz, George. **The Emergence of Modern Universities in France, 1863-1914.** Princeton:
Princeton University Press, 1983.

Presents a study of the educational reforms
in France prior to World War I. Argues that par-
tial autonomy allowed universities to meet the
needs of expanding industries by conducting research and teaching in areas outside of the regu-
lar science faculties. Shows how this led to the
creation of separate institutes for the training
of engineers.

520. Wickenden, William E. **A Comparative Study of Engineering Education in the United States and
Europe.** Lancaster, Pa.: Society for the Promotion of Engineering Education, 1929.

Presents a classic survey of late 19th and
early 20th century engineering education. Discusses the role of science in engineering.

521. Wood, Ethel Mary. **A History of the Polytechnic.**
London: Macdonald, 1965.

Provides a history of the London Polytechnic
Institute from its founding in 1838.

See also: 33, 50, 83, 86, 87, 127, 137, 292, 298, 300,
306, 312, 323, 325, 348, 350, 379, 392, 736, 737, 1141.

**SOCIETIES, RESEARCH INSTITUTES AND PROFESSIONAL
ORGANIZATIONS**

522. Allan, D.G.C. "The Society of Arts and Government, 1754-1800: Public Encouragement of

Arts, Manufactures, and Commerce in 18th Century England." **Eighteenth-Century Studies** 7 (1973-74): 434-52.

523. Allibone, T.E. "The Club of the Royal College of Physicians, the Smeatonian Society of Civil Engineers and Their Relationship to the Royal Society Club." **Notes and Records of the Royal Society of London** 22 (1967): 186-192.

Describes the early history of one of the oldest professional engineering societies.

524. Barnaby, K.C. **The Institution of Naval Architects, 1860-1960**. London: Royal Institution of Naval Architects, 1960.

Provides a history of the Institution of Naval Architects from its founding in 1860. Includes summaries of the papers published in their **Transactions**.

525. Bates, Ralph S. **Scientific Societies in the United States**. 3rd edition. Cambridge, Mass.: M.I.T. Press, 1965.

Useful as a reference tool. Describes the wide extent of scientific and technological societies in the United States.

526. Berth, Donald F. "ASME at 100: A Reflection on the Founding and a Favorite Founder." **Engineering: Cornell Quarterly** 14 (Autumn 1979): 3-11.

Focuses on the role of Robert Thurston as first president of the American Society of Mechanical Engineers and his role in engineering education at Cornell.

527. Boehm, George A.W., and Alex Groner. **Science in the Service of Mankind: The Battelle Story**. Lexington, Mass.: Lexington Books, 1972.

Provides a history of the Battelle Memorial Institute in Columbus, Ohio and its application of science to technology.

528. Buchanan, R.A. "Gentlemen Engineers: The Making of a Profession." **Victorian Studies** 26 (1983): 407-29.

Discusses the rise of the professionalization of engineering during the 19th century.

529.   _____. "Institutional Proliferation in the British Engineering Profession, 1847-1914." **Economic History Review** 38 (1985): 42-60.

Discusses the development of engineering organizations, institutions and societies between 1847 and 1914.

530.   Burke, John G. "Bursting Boilers and the Federal Power." **Technology and Culture** 7 (1966): 1-23.

Discusses how the Franklin Institute played a role in investigating the problem of bursting boilers on steam boats through a series of scientific experiments. Shows how the knowledge developed from those experiments influenced the framing of federal regulations.

531.   Cahan, David L. "The Physikalisch-Technische Reichsanstalt: A Study in the Relations of Science, Technology, and Industry in Imperial Germany." Ph.D. dissertation. Johns Hopkins University, 1980.

Describes the origins, purposes and structure of the PTR. Argues that its establishment in 1887 was due to Werner Siemens who saw it as a national institute for conducting pure science and technical research for the advancement of both science and industry. Concludes that the needs of German industry dominated its program.

532.   Calhoun, Daniel H. **The American Civil Engineering: Origins and Conflict.** Cambridge, Mass.: M.I.T. Press, 1960.

Discusses the history of American civil engineering as a profession until 1840. Analyzes the role of engineering schools in the professionalization process.

533.   Calvert, Monte. **The Mechanical Engineer in America, 1830-1910.** Baltimore: Johns Hopkins University Press, 1967.

Presents a history of the professionalization
of mechanical engineering in America. Shows that
two opposing cultures or schools of thought
played important roles in the process--shop culture and school culture. Argues that school culture sought to replace intuition and skill with
pure science and higher mathematics.

534.   Carter, Edward C. "Benjamin Henry Latrobe,
       'Learned Engineer,' the American Philosophi-
       cal Society, and the Promotion of Useful
       Knowledge and Works, 1798-1809." **Science and
       Society in Early America: Essays in Honor of
       Whitfield J. Bell, Jr.** Edited by Randolph
       Shipley Klein. Philadelphia: American Philo-
       sophical Society, 1986.

535.   Clarke, J.F. **A Century of Service to Engineering
       and Shipbuilding: A Centenary History of the
       North East Coast Institution of Engineering
       and Shipbuilders 1884-1984.** Newcastle upon
       Tyne: North East Coast Institution of Engi-
       neers and Shipbuilders, 1984.

       Provides a scholarly study of a significant
local engineering society. Discusses conflicts
within the society between university trained engineers and the more practically trained shipyard
engineers.

536.   Comune di Bolgna, ed. **Macchine scuola industria
       dal mestiere a la professionalita.** Milan:
       Bologna Societa Editrice il Mulino, 1980.

       Describes an exhibition celebrating the cen-
tenary of the Aldini-Valeriani Technical-Indus-
trial Institute of Bologna. Describes how Luigi
Aldini collected scientific and technological information from London, Manchester, Birmingham,
Glasgow, and Edinburgh and created a mechanical
physics for workmen. Traces the Aldini Laboratory from its role as a calculator and measurer
of machine performance to an institute for training pupils.

537.   Dennis, T.L., ed. **Engineering Societies in the
       Life of a Country: A Series of Lectures Com-
       memorating the 150th Anniversary of the Foun-
       dation of the Institution of Civil Engineers.**
       London: Institution of Civil Engineers, 1968.

Includes papers on engineering societies in the United States and on the Continent, as well as in Great Britain.

538.   Donkin, S.B.   "The Society of Civil Engineers (Smeatonians)." **Transactions of the Newcomen Society** 17 (1936-37): 51-71.

Investigates the history of one of the oldest professional engineering societies, founded in 1771 and known, after the death of John Smeaton, as the Smeatonians.

539.   Farrar, W.V.   "The Society for the Promotion of Scientific Industry, 1872-1876." **Annals of Science** 29 (June 1972): 81-86.

540.   Fox, Robert, and George Weisz, eds. **The Organization of Science and Technology in France, 1808-1914**. New York: Cambridge University Press, 1980.

Argues that the organization of French science and technology was less inadequate than critics have thought. Argues the France's scientific and technological educational system was similar to Germany's by 1914 but that France did not have enough industrial research labs to absorb the increased numbers of science and engineering students produced during the Third Republic. Most of the papers are quantitative.

541.   Gispen, Cornelius W.R.   "Selbstverständnis und Professionalisierung deutscher Ingenieure: Eine Analyse der Nachrufe." **Technikgeschichte** 50 (1983): 34-61.

Analyzes obituaries from the **Zeitschrift des VDI**. Discusses the history of the engineering profession in 19th-century Germany.

542.   Hahn, Roger. **The Anatomy of a Scientific Institution: The Paris Academy of Sciences, 1666-1803**. Berkeley: University of California Press, 1971.

Presents a well documented institutional history of the Academy of Sciences from its founding through the period of the French Revolution. Analyzes the role of the Academy in evaluating technology. Investigates the relationship be-

tween science and technology, particularly during
the years of the Revolution. Provides material
concerning questions about the utility of science
and the social relationship between scientists
and artisans (engineers).

543.   Hall, A. Rupert. "The Royal Society of Arts: Two
       Centuries of Progress in Science and Technol-
       ogy." **Journal of the Royal Society of Arts**
       122 (1974): 641-58.

       Discusses the role of the Royal Society of
       Arts in promoting the application of science to
       useful purposes.

544.   Hudson, Derek, and Kenneth W. Luckhurst. **The**
       Royal Society of Arts, 1754-1954. London:
       Royal Society of Arts, 1954.

       Provides a history of an institution which
       played an important role in the dissemination of
       technological knowledge.

545.   Jones, Bence. **The Royal Institution: Its**
       Founder and Its First Professors. New York:
       Arno Press, 1975.

       Reprint of the 1871 edition. Provides a his-
       tory of the early days of the Royal Institution.
       Discusses Count Rumford's practical studies on
       heat.

546.   Kelves, Daniel J. "Federal Legislation for Engi-
       neering Experiment Stations: The Episode of
       World War I." **Technology and Culture** 12
       (1971): 182-89.

       Discusses the attempt to establish federally
       funded engineering experiment stations. Dis-
       cusses the political reasons for their failure.

547.   Layton, Edwin T., Jr. **The Revolt of the Engi-**
       neers: Social Responsibility and the Ameri-
       can Engineering Profession. Cleveland: Case
       Western University Press, 1971.

       Investigates the role of science in the pro-
       fessionalization of engineers. Argues that engi-
       neers are pulled between the conflicting values
       associated with science and with business.

Reprinted in 1986 by Johns Hopkins Press, with a new introduction by the author.

548. Lea, Frederick M. **Science and Building: A History of the Building Research Station.** London: H.M.S.O., 1971.

Provides a history of the research carried out at the Building Research Station and how it contributed to engineering practice.

549. Lewis, Robert A. "Government and the Technological Sciences in the Soviet Union: The Rise of the Academy of Sciences." **Minerva** 15 (1977): 174-99.

Discusses the role of the Academy of Sciences in developing technological sciences.

550. Ludwig, Karl Heinz, and Wolfgang König, eds. **Technik, Ingenieure und Gesellschaft: Geschichte des Vereins Deutscher Ingenieure 1856-1981.** Dusseldorf: VDI Verlag, 1981.

Presents a history of the VDI (the Association of German Engineers) which played a significant role in the establishment of technical universities and the introduction of the laboratory approach to engineering. Discusses the decline of the organization during the Nazi era and its revival in the post-war period.

551. MacLeod, Roy M., and E. Kay Andrews. "Scientific Advice in the War at Sea, 1915-1917: The Board of Invention and Research." **Journal of Contemporary History** 6 (2) (1971): 3-40.

Describes the impact of science-based technology on military techniques and tactics. Chronicles the resistance by the military to the rise of scientists, technologists and inventors to positions of importance.

552. McMahon, A. Michael. "'Bright Science' and the Mechanic Arts: The Franklin Institute and Science in Industrial America, 1824-1976." **Pennsylvania History** 47 (1980): 351-68.

Discusses the interaction of science and technology in the Franklin Institute.

553. Makepeace, Chris E. **Science and Technology in Manchester: Two Hundred Years of the Lit. and Phil.** Manchester: Manchester Literary and Philosophical Publications, 1984.

Describes the role of the members of the Literary and Philosophical Society to the industrial development of Manchester. Brief account with many illustrations.

554. Martin, Thomas. "Early Years at the Royal Institution." **British Journal for the History of Science** 2 (1964): 99-115.

Discusses the Royal Institution during the period 1800 to 1802.

555. _____. "Origins of the Royal Institution." **British Journal for the History of Science** 1 (1962): 49-63.

Discusses the founding of the Royal Institution.

556. Merritt, Raymond. **Engineering in American Society, 1850-1875.** Lexington: University Press of Kentucky, 1969.

Focuses on the professionalization of civil engineering in America. Argues that professionalization was not based on esoteric know-how but on new education that stressed practicality. Ignores the role of science and neglects such schools as M.I.T. Focuses only on Rensselaer Polytechnic Institute and Stevens Institute of Technology.

557. Middleton, W.E. Knowles. **Mechanical Engineering at the National Research Council of Canada, 1929-1951.** Waterloo, Ont.: Wilfred Laurier University Press, 1984.

Discusses the connection between the NRC and the Royal Canadian Air Force. Describes research done on problems of flying in cold climates, such as carburetor icing, hydraulic fluids in cold temperature, and de-icing of wings and propellers. Argues that the branch-plant nature of Canadian industry gave the NRC few private clients since research was done in U.S. labs. Shows how the NRC turned to the military in World

War II. Argues that the work of the NRC led to a
post-war effort to establish a domestic aircraft
industry.

558.   Morgan, Arthur E. **Dams and Other Disasters: A
        Century of the Army Corps of Engineers.**
        Boston: Porter Argent, 1971.

        Surveys the work of the Army Corps of Engi-
        neers. Argues that the Corps failed to develop
        inclusive engineering analysis. Provides evi-
        dence that blind loyalty within the Corps led to
        an attempt to stop construction of the Eads
        bridge in St. Louis, and for similar reasons the
        Corps supported the use of only levees for flood
        control in the Mississippi valley. Notes that
        the Corps opposed the use of hydraulic laborato-
        ries.

559.   Morrell, Jack, and Arnold Thackray. **Gentlemen of
        Science: Early Years of the British Associa-
        tion for the Advancement of Science.** Oxford:
        Clarendon Press, 1981.

        Contains a section on the ideologies of sci-
        ence and the utilities of science. Includes a
        section on the role of mechanical science within
        the association. Discusses the interaction of
        science and technology within the association.
        Argues that the practical arts were seen as sub-
        ordinate to theoretical science.

560.   Mount, Ellis. **Ahead of Its Time: The Engineer-
        ing Societies Library, 1913-80.** Hamden,
        Conn.: Linnet Books, 1982.

        Provides a survey of the history and current
        activities of the Engineering Societies Library,
        which was created through the merger of the li-
        braries of the American Society of Mechanical En-
        gineers, the American Institute of Mining Engi-
        neers, the American Institute of Electrical Engi-
        neers, and later the American Society of Civil
        Engineers.

561.   Ornstein, Martha. **The Role of Scientific Soci-
        eties in the Seventeenth Century.** 2nd edi-
        tion. Hamden, Conn.: Shoe String Press,
        1963.

Reprint of the 1928 edition. Detailed study
of the rise of scientific societies in the 17th
century including the Royal Society of London and
the Academie des Sciences in Paris. Includes a
discussion of how the societies encouraged the
application of science to practice.

562.   Parsons, R.H. **A History of the Institution of
Mechanical Engineers 1847-1947**. London: Institution of Mechanical Engineers, 1947.

Discusses the role of the Institution in
stimulating research into engineering theory
through meetings and publications.

563.   Pursell, Carroll W. "The Technical Society of
the Pacific Coast, 1884-1914." **Technology
and Culture** 17 (1976): 702-716.

Argues that engineering societies played a
critical role in linking science and technology
in 19th-century America. Traces the history of
the Technical Society of the Pacific Coast. Concludes that the Society was a mirror image twin
of the scientific societies (see item 202).

564.   Purver, Margery. **The Royal Society: Concept and
Creation**. Introduction by H.R. Trevor-Roper.
Cambridge, Mass.: M.I.T. Press, 1967.

Rejects some standard versions of the founding of the Royal Society and argues that it grew
out of the experimental science club at Oxford
and that it was founded to carry out Bacon's
ideal of Salomon's House that he presented in the
New Atlantis. Concludes that the Royal Society's
purpose was to advance the mechanical arts
through experiments in technology.

565.   Pyatt, Edward. **The National Physics Laboratory:
A History**. Bristol: Adam Hilger, 1983.

Traces the history of the National Physics
Laboratory from its founding in 1899 as an institution to provide scientific research and standards for technological development. Gives both
a technical and administrative history. Includes
descriptions of work on aerodynamics, stress, and
ship design.

566. Ritchie-Calder, Peter. "The Lunar Society of
      Birmingham." **Scientific American** 6 (1982):
      136-45.

          Discusses a group of inventors and scientists
      (including Watt and Priestley) who helped to
      transform science and technology in 18th-century
      Britain.

567. Rolt, L.T.C. **The Mechanicals: Progress of a
      Profession.** London: Heinemann, 1967.

          Provides a history of the Institution of Me-
      chanical Engineers. Focuses on an analysis of
      papers on prime movers given at the Institution.
      Emphasizes technical problems more than social
      history. Lacks a discussion of the Institution's
      role in engineering education.

568. Schofield, Robert E. "Histories of Scientific
      Societies: Needs and Opportunities for Re-
      search." **History of Science** 2 (1963): 70-
      83.

          Focuses on British scientific societies from
      the 17th to the 19th centuries. Notes that many
      societies included engineers as members.

569. _____. "The Industrial Orientation of Science
      in the Lunar Society of Birmingham." **Isis**
      48 (1957): 408-15.

          Discusses the relationship between science
      and technology among an influential group of sci-
      entists such as Joseph Priestley, and industrial-
      ists, such as James Watt, who made up the Lunar
      Society. Argues that their view of science was
      oriented toward practical applications.

570. _____. **The Lunar Society of Birmingham: A So-
      cial History of Provincial Science and Indus-
      try in Eighteenth-Century England.** London
      and New York: Oxford University Press, 1963.

          Detailed description of the interaction be-
      tween science and technology among the members of
      the Lunar Society, which included James Watt,
      Josiah Wedgwood, Matthew Boulton and Joseph
      Priestley. Important study of the social basis
      of the interaction between science and technol-
      ogy.

571.   Seely, Bruce E.   "The Scientific Mystique in En-
       gineering: Highway Research at the Bureau of
       Public Roads, 1918-1940." **Technology and
       Culture**   25 (1984): 798-831.

       Analyzes the role of the Bureau of Public
       Roads in the development of a scientific approach
       to engineering. Argues that the application of
       science in some areas of engineering was more
       cautious than has been expected. Shows that the
       Bureau of Public Roads' infatuation with experi-
       mental methods, usually associated with science,
       hindered the solution of practical engineering
       problems and failed to enhance a theoretical un-
       derstanding of engineering problems.

572.   Shinn, Terry.   "The Genesis of French Industrial
       Research, 1880-1940." **Social Science Infor-
       mation**   19 (1980): 609-40.

573.   Sinclair, Bruce.   **Early Research at the Franklin
       Institute: The Investigation into the Causes
       of Steam Boiler Explosions, 1830-1837.**
       Philadelphia: Franklin Institute, 1966.

       Describes the application of scientific theo-
       ries and methods at the Franklin Institute to the
       study of the cause of steam boiler explosions.

574.   _____.   **Philadelphia's Philosopher Mechanics: A
       History of the Franklin Institute, 1824-1865.**
       Baltimore: Johns Hopkins University Press,
       1975.

       Describes the social, intellectual, politi-
       cal, economic and technical forces that shaped
       the early years of the Franklin Institute. Shows
       how a focus on experimental research emerged un-
       der Alexander Dallas Bache. Details the interac-
       tion of science and technology within the Insti-
       tute. Important work.

575.   _____.   "Science, Technology, and the Franklin
       Institute."   **The Pursuit of Knowledge in the
       Early American Republic.**   Edited by A. Oleson
       and S.C. Brown.   Baltimore: Johns Hopkins
       University Press, 1976, pp. 194-207.

       Discusses the interaction of science and
       technology in the Franklin Institute.

576. _____, and James F. Hull. **A Centennial History
      of the American Society of Mechanical Engi-
      neers, 1880-1980.** Toronto: University of
      Toronto Press, for the American Society of
      Mechanical Engineers, 1980.

      Discusses the establishment and the history
      of the ASME. Includes a discussion of the role
      of Robert Thurston as first president of the So-
      ciety.

577. Ulrich, Lars. **Ingenieure in der Frühindustrial-
      isierung. Staatliche und private Techniker
      im Königreich Hannover und an der Ruhr (1815-
      1873).** Göttingen: Vandenhoeck & Ruprecht,
      1978.

      Describes the role of engineers as civil ser-
      vants in Lower Saxony and in engineering compa-
      nies in the Ruhr. Argues that knowledge of sci-
      entific principles increased engineers' status.
      Illustrates how technicians were replaced by sci-
      entifically trained engineers. Shows that these
      engineers became more than higher level artisans.

578. Vernon, K.D.C. "The Foundation and Early Years
      of the Royal Institution." **Proceedings of
      the Royal Institution of Great Britain** 39
      (4) (1963): 364-402.

      Discusses the early years (1799-1803) of the
      Royal Institution, whose role it was to apply
      science to useful purposes.

579. Wright, Esther Clark. "The Early Smeatonians."
      Transactions of the Newcomen Society 18
      (1937-38): 105-110.

      Provides a history of one of the first pro-
      fessional engineering societies founded in 1771
      as the Society of Civil Engineers and known after
      the death of John Smeaton as the Smeatonians.

See also: 14, 106, 126, 287, 318, 322, 647, 648, 1437,
1487.

CHAPTER V: THE PRE-HISTORY OF ENGINEERING SCIENCE

580. Anderson, R.C. "Early Books on Shipbuilding and Rigging." **Mariner's Mirror** 10 (January 1924): 53-64.

   Focuses on books on shipbuilding published from 1536 to 1720.

581. Ariotti, Piero E. "Huygens: Aviation Pioneer Extraordinary." **Annals of Science** 36 (1979): 611-24.

   Discusses Huygens' interest in and contribution to the theory of aviation.

582. Bayerl, Gunter. Technische Intelligenz im Zeitalter der Renaissance." **Technikgeschichte** 45 (1978): 336-53.

   Investigates technical knowledge in the Renaissance.

583. Bedini, Silvio A. "The Role of Automata in the History of Technology." **Technology and Culture** 5 (1964): 24-42.

   Argues that the scientific writings of Ktesibius, Philon, and Heron influenced the development of 16th- and 17th-century automata. Concludes that the attempt to recreate life by mechanical means played a significant role in the history of technology.

584. Bennett, J.A. "The Mechanics' Philosophy and the Mechanical Philosophy." **History of Science** 24 (1986): 1-28.

   Discusses the role of practical science and scientific instruments in the development of the mechanical philosophy.

585.   _____. "Robert Hooke as Mechanic and Natural
        Philosopher." **Notes and Records of the Royal
        Society of London** 35 (1980): 33-48.

        Discusses the relationship between science
        and technology in the work of Robert Hooke.

586.   Burke, John G., ed. **The Uses of Science in the
        Age of Newton.** Berkeley: University of
        California Press, 1983.

        Contains several papers on the application of
        science to technology. Includes items 611, 614,
        682.

587.   Clagett, Marshall. "Leonardo da Vinci and the
        Medieval Archimedes." **Physis** 11 (1969):
        100-151.

        Analyzes the role of Archimedes' theories in
        the scientific and technical works of Leonardo.
        Argues that Leonardo's knowledge of Archimedes
        came from medieval sources.

588.   _____. **The Science of Mechanics in the Middle
        Ages.** Madison: University of Wisconsin
        Press, 1959.

        Discusses the relationship between the sci-
        ence of mechanics and the five simple machines of
        the Greeks, but the work also shows that most me-
        dieval mechanical technology was completely sepa-
        rate from the science of mechanics.

589.   Crombie, Alistair C. **Medieval and Early Modern
        Science.** 2 volumes. Garden City: Doubleday,
        1959.

        Surveys the history of science during the me-
        dieval and early modern periods. Includes a sec-
        tion on "Technics and Science in the Middle
        Ages." Earlier version published under the title
        **Augustine to Galileo: The History of Science
        A.D. 400-1650.**

590.   Daujat, Jean. "Note sur les origines de
        l'hydrostatique." **Thales** 1 (1934): 53-56.

591. Davis, Richard Beale. "Science and Technology,
      Including Agriculture." **Intellectual Life in
      the Colonial South, 1585-1763.** Knoxville:
      The University of Tennessee Press, 1978.

      Argues that the social, environmental and
      economic background of the Southern colonies made
      technology less dependent upon European ideas,
      while science was more dependent.

592. Dee,John. **The Mathematical Preface to Euclid
      (1570).** Introduction by Allen G. Debus. New
      York: Science History Publications, 1975.

      Presents Dee's preface to Henry Billingsley's
      first English translation of Euclid. Describes
      the application of geometry to such technological
      problems as navigation, mechanics, architecture
      and hydraulics.

593. Dorn, Harold, and Robert Mark. "The Architecture
      of Christopher Wren." **Scientific American**
      245 (January 1981): 160-73.

      Argues that Wren did not use his knowledge of
      theoretical mechanics in designing his buildings.

594. Drachmann, A.G. **Ktesibios, Philon and Heron: A
      Study in Ancient Pneumatics.** Copenhagen:
      Ejnar Munksgaard, 1948.

      Analyzes the automata of the Helenistic pe-
      riod and their links to pneumatic theory.

595. _____. **The Mechanical Technology of Greek and
      Roman Antiquity.** Madison: University of
      Wisconsin Press, 1962.

      Covers the works of Ktesibios, Archimedes,
      Vitruvius, and Heron of Alexandria. Concludes
      that ancient engineers were able to apply theory
      to practice and that some engines designed on pa-
      per were forms of applied mechanics.

596. Drake, Stillman, and I.E. Drabkin, eds. **Mechan-
      ics in Sixteenth-Century Italy.** Madison:
      University of Wisconsin Press, 1969.

      Provides selections from the works of
      Tartaglia, Benedetti, Guido Ubaldo, and Galileo.
      Includes works on strength of materials and the

mechanics of simple machines. Most selections
are annotated by the editors. Bibliography includes a section on the technological tradition.

597. Ferguson, Eugene S. "Leupold's Theatrum Machi-
      narum: A Need and an Opportunity." **Technol-
      ogy and Culture** 12 (1971): 64-68.

      Points out the importance of Jacob Leupold's
      18th-century work on machines. Notes that it is
      one of the earliest works to treat machines ana-
      lytically. Argues that Leupold focused on the
      elements that went into machines.

598. _____, ed. **The Various and Ingenious Machines
      of Agostino Ramelli.** Translated by M.T.
      Gnudi. Baltimore: Johns Hopkins University
      Press, 1976.

      Contains almost 195 plates of 17th-century
      technology. Provides an early example of a sys-
      tematic study of machines. Contains a biographi-
      cal study of Ramelli and an essay on his influ-
      ence on later generations.

599. Filarete, Antonio Vaverlino. **Treatise on Archi-
      tecture.** 2 volumes. Translated and intro-
      duction by John R. Spencer. New Haven: Yale
      University Press, 1965.

      Translation of an important study on Renais-
      sance building and architecture.

600. Foley, Vernard. "Leonardo's Contributions to
      Theoretical Mechanics." **Scientific American**
      225 (September 1986): 108-13.

601. Galilei, Galileo. **Two New Sciences.** Translated
      by Henry Crew and Alfonso de Salvio. Intro-
      duction by Antonio Favaro. New York: Dover
      Publications, 1954.

      Modern edition and translation of Galileo's
      important work which includes a discussion of the
      strength of materials and the motion of projec-
      tiles. Based on the 1638 Leyden edition.

602. Garcia-Diego, J.A. "The Chapter on Weirs in the
      Codex of Juanelo Turriano." **Technology and
      Culture** 17 (1976): 217-234.

Discusses a sixteenth century work on hydraulic engineering. Describes the wooden,
stone, masonary and earthen weirs presented in
the codex. Shows that the weirs reflected technical skills that were based on experiments.
Concludes that the skills needed were superior to
those of Turriano and his contemporaries.

603. Gibbs-Smith, Charles H. **The Inventions of
Leonardo da Vinci.** New York: Scribner's
Sons, 1978.

Describes, with text and illustrations, the
mechanical inventions of Leonardo. Includes his
work on aeronautics, weapons, machines, naval
crafts, land vehicles, anatomy, natural phenomena, and architecture.

604. _____. **Leonardo de Vinci's Aeronautics.**
London: H.M.S.O., 1967.

Describes Leonardo's investigations into the
theory of flight.

605. Gille, Bertrand. **Engineers of the Renaissance.**
Cambridge, Mass.: M.I.T. Press, 1966.

Provides a survey of engineering during the
Renaissance. Includes a discussion of the application of science to technology.

606. _____. **Les Mécaniciens grecs: La naissance de
la technologie.** Paris: Editions du Seuil,
1980.

Argues that a "system of technology" and a
"system of science" interacted in classical
Greece to form a new period of progress in technology. Argues that the School of Alexandria influenced the building of fortifications, harbors,
machines of war, water clocks, and ships. Shows
that these works stimulated the physicists of the
time. Discusses the role of Archimedes. Concludes that science and technology flourished
only through their interactions.

607. Goldbeck, Gustav. "Die Feuremaschine von
Leonardo da Vinci und ihre Deutungen." **Technikgeschichte** 39 (1972): 1-10.

608. Guericke, Otto von. **Experimenta nova.** Aalen,
       Germany: O. Zellar, 1962.

       Facsimile edition of the 1672 work in which
       von Guericke describes his experiment with the
       Magdeburg sphere which showed the power of the
       atmosphere and led to the invention of the steam
       engine.

609. Hacker, Barton C. "Greek Catapults and Catapult
       Technology: Science, Technology, and War in
       the Ancient World." **Technology and Culture**
       9 (1968): 34-50.

       Argues that Greek mechanicians such as Heron
       of Alexandria and Philon of Byzantium, in con-
       trast to Greek scientists, showed an interest in
       practical applications of their work. Concludes
       that mechanicians made catapult construction sys-
       tematic.

610. Hall, A. Rupert. **Ballistics in the Seventeenth
       Century.** Cambridge: Cambridge University
       Press, 1952.

       Analyzes the study of ballistics in the 17th
       century, including the influence of science on
       the theory of ballistics. Discusses the contri-
       butions of Galileo, Mersenne, Huygens, Newton,
       Leibniz, and Johann Bernoulli.

611. _____. "Gunnery, Science, and the Royal Soci-
       ety." **The Uses of Science in the Age of
       Newton** (see item 586), pp. 111-141.

       Analyzes approaches to ballistics in the 17th
       century, including those by members of the Royal
       Society. Concludes that scientists wanted to
       show the utility of science, but distinguishes
       mathematical idealists from those more practical.
       Believes that the interest of members of the
       Royal Society in ballistics was exaggerated.

612. _____. "Science, Technology and Warfare, 1400-
       1700." **Science, Technology and Warfare.**
       Edited by M.D. Wright and L.J. Paszek.
       Washington, D.C., 1970.

       Argues that during this period there was lit-
       tle connection between science and technology in
       the area of warfare.

613.   Hall, Bert S., and Delmo C. West, eds. **On Pre-Modern Technology and Science: A Volume of Studies in Honor of Lynn White, Jr.** Malibu: Undena Publications, 1976.

614.   Hall, Marie Boas. "Oldenburg, the **Philosophical Transactions**, and Technology." **The Uses of Science in the Age of Newton** (see item 586), pp. 21-47.

       Discusses the Royal Society's attempt to compile the histories of trades. Concludes that Oldenburg and other members of the Royal Society became disenchanted with the project.

615.   _____, ed. **The "Pneumatics" of Hero of Alexandria: A Facsimile of the 1851 Woodcroft Edition.** New York: American Elsevier Publishing Co., 1973.

       Presents a 19th-century translation of the mechanical, hydraulic and pneumatic works of the first-century Greek mechanician, Heron of Alexandria. Provides an example of the application of a modified version of the Aristotelian theory of matter to technology. Provides details concerning Bennet Woodcroft, who was a professor of machinery, and his interest in translating the work.

616.   Hart, Clive. **The Dream of Flight: Aeronautics from Classical Times to the Renaissance.** London: Faber, 1972.

       Traces the pre-history of aeronautical theory.

617.   Hart, Ivor B. **Mechanical Investigations of Leonardo de Vinci.** Introduction by Ernest A. Moody. Berkeley: University of California Press, 1963.

       Provides a revised and corrected version of the 1925 edition. Discusses Leonardo's mechanical experimentations and their contribution to his technology. For a later version see item 618.

618.   _____. **The World of Leonardo da Vinci, Man of Science, Engineer and Dreamer of Flight.** New York: Viking Press, 1962.

Discusses the relationship between science, experimentation and technology in the work of Leonardo. Special emphasis is placed on Leonardo's interest in the science of flight and his attempt to design artificial flying machines.

619.   Heydenreich, Ludwig, and Bern Dibner, and Ladislao Reti. **Leonardo the Inventor.** London: Hutchinson, 1981.

Describes the contributions of Leonardo to the theory of technology.

620.   Hill, Donald R. **A History of Engineering in Medieval and Classical Times.** La Salle, Ill.: Open Court Publishing, 1984.

Especially useful for its discussion of the Islamic period. Includes some material on the relationship between scientific theory and the development of automata.

621.   _____, ed. **The Book of Knowledge of Ingenious Mechanical Devices by Ibn al-Razzaz al-Jazari.** Dordrecht and Boston: D. Reidel, 1974.

Presents an annotated translation of the 13th-century Arabic treatise on water clocks, automata, water raisers, trick drinking vessels and other mechanisms. Traces the tradition of such inventions back to Heron of Alexandria. Provides a key example of an early attempt to systematically present and study hydraulic machines and mechanisms.

622.   Hollister-Short, G. "The Sector and Chain: An Historical Enquiry." **History of Technology** 4 (1979): 149-85.

Traces the history of the sector and chain (or rope) mechanism from Leonardo to Newcomen and Smeaton. Includes a discussion of the works of Huygens, Leupold, La Hire, and Belidor.

623.   Hulme, W. Wyndham. "Introduction to the Litera-
        ture of Historical Engineering to the Year
        1640." **Transactions of the Newcomen Society**
        1 (1920-21): 7-15.

        Surveys early books on technology.

624.   Jenson, Martin. **Civil Engineering Around 1700**.
        Copenhagen: Danish Technical Press, 1969.

        Contains extracts and illustrations from
        17th- and 18th-century engineering texts, includ-
        ing Leupold's and Belidor's.

625.   Keller, Alexander G. "Archimedean Hydrostatic
        Theorems and Salvage Operations in 16th-Cen-
        tury Venice." **Technology and Culture** 12
        (1971): 602-17.

        Shows how mathematical theorems were applied
        to practical problems in Renaissance Italy. Ar-
        gues that theoria and praxis were brought to-
        gether by the assumption that secular problems
        could be solved by ancient wisdom.

626.   _____. "A Manuscript Version of Jacques
        Besson's Book of Machines, with His Unpub-
        lished Principles of Mechanics." **On Pre-Mod-
        ern Technology and Science: A Volume of Stud-
        ies in Honor of Lynn White, Jr.** (item 613),
        pp. 75-103.

627.   _____. "Mathematical Technologies and the
        Growth of the Idea of Technical Progress in
        the 16th Century." **Science, Medicine and So-
        ciety in the Renaissance.** Edited by Allen
        Debus. New York: Science History, 1972, vol-
        ume 1, pp. 11-27.

        Investigates the relationship between mathe-
        matics and technology in the Renaissance.

628.   _____. Mathematicians, Mechanics and Experimen-
        tal Machines in Northern Italy in the 16th
        Century." **The Emergence of Science in West-
        ern Europe.** Edited by M. Crosland. London:
        Macmillan, 1976, pp. 15-34.

        Shows how Renaissance engineers developed
        mathematical theories of machines.

629.   _____. "Mathematics, Mechanics, and the Origins
        of the Culture of Mechanical Invention."
        Minerva 23 (1985): 348-61.

        Examines the relationship between science and
        technology in the Renaissance by focusing on the
        relationship between mathematics and the arts.

630.   _____. "The Missing Years of Jacques Besson,
        Inventor of Machines, Teacher of Mathematics,
        Distiller of Oils, and Huguenot Pastor."
        Technology and Culture 14 (1973): 28-39.

        Presents details which fill in the back-
        ground of the life of the 16th-century inventor,
        Besson who authored **Théâtre des instruments mathé-
        matiques et mécaniques**, which was one of the
        first printed books to deal with mechanical in-
        ventions. Speculates that his mechanical ideas
        may have been influenced by his "missing year" in
        Switzerland.

631.   _____. "Pneumatics, Automata and the Vacuum in
        the Work of Giambattista Aleotti." **British
        Journal for the History of Science** 3 (1967):
        338-47.

        Discusses the role of mathematics and science
        in the development of Renaissance technology.

632.   _____. "A Renaissance Humanist Looks at 'New'
        Inventions: The Article 'Horologium' in
        Giovanni Tortelli's **De orthographia**." **Tech-
        nology and Culture** 11 (1970): 345-65.

        Describes and translates a 15th-century list
        of inventions. Concludes that Renaissance human-
        ists believed that such lists demonstrated that
        the arts progressed.

633.   _____. "Renaissance Waterworks and Hydrodynam-
        ics." **Endeavour** 25 (September 1966): 141-
        145.

        Discusses the interest in hydraulic machinery
        during the 16th and 17th centuries. Concludes
        that this interest led to a wider understanding
        of hydraulic phenomena.

634. _____. **A Theatre of Machines.** London: Chapman
    & Hall, 1964.

    Includes illustrations of machines from the
    16th- and 17th-century works of Ramelli, Besson,
    Strada, Bronca and Zonca. Includes a brief in-
    troduction. Gives some idea of the early inter-
    est in a systematic study of machines.

635. Kenny, Andre. "Vacuum Theory and Technique in
    Greek Science." **Transactions of the Newcomen
    Society** 37 (1964-65): 47-55.

    Describes the relationship between theory of
    the vacuum and the development of pumps in an-
    cient Greece.

636. Kilgour, Frederick G. "Vitruvius and the Early
    History of Wave Theory." **Technology and Cul-
    ture** 4 (1963): 282-86.

    Discusses the ancient Stoic theory of sound
    and its role in Vitruvian architecture. Con-
    cludes that engineers played a major role in tak-
    ing over and developing a wave theory of sound.

637. Klemm, Friedrich. "Die Rolle der Technik in der
    italienischen Renaissance." **Tech-
    nikgeschichte** 32 (1965): 221-43.

    Investigates the role of technology in the
    Italian Renaissance.

638. Knobloch, Eberhard. "Mariano di Jacopo detto
    Taccolas **De machinis:** Ein Werk der italienis-
    chen Fruhrenaissance." **Technikgeschichte** 48
    (1981): 1-27.

    Analyzes Taccola's book on machines.

639. Kyeser, Conrad. **Bellifortis.** Volume I: **Facsim-
    ile--Ausgabe der Pergamenthandshrift, Cod.
    Ms. philos. 63 der Universitätsbibliothek,
    Göttingen.** Volume II: **Umschrift und Uber-
    setzung nebst Erlauterungen.** Edited by Gotz
    Quarg. Dusseldorf: Verlag des Vereins
    Deutscher Ingenieure, 1967.

    Facsimile of a 1405 manuscript which is the
    first technological treatise of the 15th century.
    Includes a discussion of magic as a part of tech-

nology. For an analysis of the work, see item
683.

640. Lane, Frederic Chapin. **Venetian Ships and Ship-
      builders of the Renaissance.** Baltimore:
      Johns Hopkins University Press, 1934.

      Discusses methods of design and ships' pro-
      portions during the Renaissance.

641. Long, Pamela O. "The Contribution of Architec-
      tural Writers to a 'Scientific' Outlook in
      the 15th and 16th Centuries." **Journal of Me-
      dieval and Renaissance Studies** 15 (1985):
      265-98.

642. Mark, Robert. **Experiments in Gothic Structure.**
      Cambridge, Maso: M.I.T. Press, 1982.

      Uses modern engineering analysis to study
      Gothic cathedrals to show that the physical ele-
      ments of the cathedral played a necessary struc-
      tural role in the building. Concludes that me-
      dieval engineers used a quasi-scientific system
      to design such structures.

643. Merton, Robert K. **Science, Technology and Soci-
      ety in Seventeenth Century England.** New
      York: Harper Torchbooks, 1970.

      Originally published in Osiris in 1938.
      Classic study on the sociology of science and
      technology. Argues that practical problems, such
      as those involving gunnery, influenced the choice
      of scientific problems for study by members of
      the Royal Society.

644. Moran, Bruce T. "German Prince-Practitioners:
      Aspects in the Development of Courtly Sci-
      ence, Technology, and Procedures in the Re-
      naissance." **Technology and Culture** 22
      (1981): 253-74.

      Argues that German prince-practitioners
      helped to bridge the gulf between scholars and
      craftsmen and helped to introduce scientific con-
      cepts into technological activity.

645. Needham, Joseph. "The Pre Natal History of the
      Steam Engine." **Transactions of the Newcomen
      Society** 35 (1962-63): 3-56.

Discusses the development of the steam jet, pumps and the theory of the vacuum in ancient Greece, medieval Europe and China. Shows how they all played a role in the development of the steam engine.

646. Norden, Marcel. "Introduction and Analysis of the 21 Books of Devices and Machines of Pseudo Juanelo Turriano." **History and Technology** 2 (1986): 336-65.

647. Ochs, Kathleen H. The Failed Revolution in Applied Science: Studies of Industry by Members of the Royal Society of London, 1660-1688." Ph.D. dissertation. University of Toronto, 1981.

Analyzes the attempt by the Royal Society to create a history of artisans' trades.

648. _____. "The Royal Society of London's History of Trades Programme: An Early Episode in Applied Science." **Notes and Records of the Royal Society of London** 39 (1985): 129-58.

649. Olson, Richard. **Science Deified and Science Defied: The Historical Significance of Science in Western Culture from the Bronze Age to the Beginnings of the Modern Era, ca. 3500 B.C. to ca. A.D. 1640.** Berkeley: University of California Press, 1982.

Discusses the impact of science on technology.

650. Olszewski, Eugeniusz. "Leonardo Da Vinci jako prekursor nauk technicznych." **Kwartalnik Historii i Techniki** 14 (1969): 603-13.

Discusses the role of Leonardo as the precursor of engineering science.

651. Parsons, William Barclay. **Engineers and Engineering in the Renaissance.** Introduction by Robert S. Woodbury. Cambridge, Mass.: M.I.T. Press, 1968.

Includes a discussion of the role of science in Renaissance engineering, especially in the work of Leonardo da Vinci.

652. Plommer, Hugh. **Vitruvius and Later Roman Build-
      ing Manuals.** New York and London: Cambridge
      University Press, 1973.

      Provides an example of an early attempt to
      systematize building practices.

653. Porta, Giambattista della. **Natural Magick.**
      Preface by Derek J. de Solla Price. New
      York: Dover, 1957.

      Reprint of the 1658 English translation from
      the 1589 original. Treats technology as a form
      of natural magic. Contains useful insights into
      the relationship between science, magic and tech-
      nology during the Renaissance.

654. Prager, Frank D. "Fontana on Fountains: Venetian
      Hydraulics of 1418." **Physis** 13 (1971): 341-
      60.

      Discusses Giovanni Fontana's contribution to
      15th-century hydraulics.

655. _____. "A Manuscript of Taccola, Quoting
      Brunelleschi, on Problems of Inventors and
      Builders." **Proceedings of the American
      Philosophical Society** 112 (1968): 131-49.

      Describes the relationship between the inven-
      tor Taccola and the architect Brunelleschi.

656. _____, and Gustina Scaglia. **Brunelleschi: Stud-
      ies of His Technology and Inventions.**
      Cambridge, Mass.: M.I.T. Press, 1970.

      Discusses Brunelleschi's role as a builder,
      especially his work on the Cathedral of Santa
      Maria del Fiore in Florence. Provides a somewhat
      flawed structural analysis of the dome of the
      Cathedral. Concludes that Brunelleschi was a
      great structural innovator.

657. _____ eds. **Mariano Taccola and His Book "De in-
      geneis."** Cambridge, Mass.: M.I.T. Press,
      1971.

      Provides an example of the systemization of
      technological knowledge through written texts
      that arose during the 15th and 16th centuries.

658.   Price, Derek J. de Solla. "Automata and the Ori-
        gins of Mechanism and Mechanistic Philoso-
        phy." **Technology and Culture** 5 (1964): 9-
        23.

        Argues that a mechanistic philosophy led to
        the development of mechanical devices such as au-
        tomata, rather than the other way around. Con-
        cludes that automata represent the triumph of ra-
        tional mechanistic explanations over vitalistic
        and theological explanations.

659.   Reti, Ladislao. "The Codex of Juanelo Turriano
        (1500-1585)." **Technology and Culture** 8
        (1967): 53-66.

        Describes the work of Turriano and his stud-
        ies of machines. Includes biographical material
        on Turriano.

660.   _____. "The Codex of Juanelo Turriano (1500-
        1585) in the Biblioteca Nacional of Madrid
        and Its Importance for the History of Tech-
        nology." **Actes XIe Congrès International
        d'Histoire des Sciences** 6 (1965): 79-83.

661.   _____. "Die wiedergefundenen Leonard-
        Manuskripte der Biblioteca Nacional in
        Madrid." **Technikgeschichte** 34 (1967): 193-
        225.

        Discusses the discovery, in Madrid, of
        Leonardo's notebooks on machines.

662.   _____. "Francesco di Giorgio Martini's Treatise
        on Engineering and Its Plagiarists." **Tech-
        nology and Culture** 4 (1963): 287-98.

        Discusses Martini's (1439-1501) Trattato di
        Architectura and its contribution to mechanical
        engineering and a study of machines. Argues that
        the work, which was plagiarized by others, influ-
        enced many, including Leonardo. Briefly dis-
        cusses and shows examples of the mechanical de-
        vices in the work.

663.   _____. "The Horizontal Waterwheels of Juanelo
        Turriano (ca. 1565): A Prelude to Bascale."
        **Actes XIIe Congrès International d'Histoire
        des Sciences** 10b (1968): 79-82.

664.   _____. "Il moto dei prioietti e del pendolo
        secondo Leonardo e Galileo." **Le Machine** 1
        (December 1968): 63-89.

        Suggests that Galileo's scientific method was
        influenced by experimental and theoretical ideas
        of Renaissance engineers.

665.   _____. "Leonardo and Ramelli." **Technology and
        Culture** 13 (1972): 577-605.

        Argues that Agostino Ramelli's ideas on tech-
        nology were influenced by Leonardo. Compares
        drawings in the **Madrid Codices** of Leonardo with
        drawings of Ramelli. Concludes that Leonardo's
        theory of machines did not exist within a vacuum
        but that his ideas were diffused through artisans
        and their assistants to other engineers.

666.   _____. "The Leonardo Da Vinci Codices in the
        Biblioteca Nacional of Madrid." **Technology
        and Culture** 8 (1967): 437-45.

        Describes the book of machines by Leonardo
        discovered in Madrid in 1965. Provides evidence
        of Leonardo's systematic study of machines.

667.   _____. "Leonardo on Bearings and Gears." **Sci-
        entific American** 224 (February 1971): 101-
        110.

        Analyzes Leonardo's investigations of fric-
        tion, bearings and gears.

668.   _____. "On the Efficiency of Early Horizontal
        Waterwheels." **Technology and Culture** 8
        (1967): 388-94.

        Analyzes sections from the Codex of Juanelo
        Turriano.

669.   _____, ed. **The Unknown Leonardo.** New York:
        McGraw-Hill Book Co., 1974.

        Large, profusely illustrated book which rein-
        terprets the work of Leonardo in light of the
        discovery of the **Madrid Codices,** which contained
        his studies of machines. Includes chapters on
        mechanics of water and stone, elements of ma-
        chines, and machines and weaponry, along with

chapters on painting and music. Reti in his
chapter on "Elements of Machines" concludes that
Leonardo was attempting a systematic study of machines and mechanisms and was a precursor of modern engineering theory.

670.   _____, and Bern Dibner. **Leonardo Da Vinci:**
        Technologist. Norwalk, Conn.: Burndy
        Library, 1969.

        Examines the role of Leonardo as an engineer.
        Argues that Leonardo was attempting a systematic
        study of machines and mechanisms.

671.   Rink, Evald. "Jacob Leupold and His **Theatrum**
        Machinarum." **Library Chronicle** 38 (Spring
        1972): 123-35.

        Documents the bibliographical history of one
        of the first systematic studies of machines.
        Also surveys the biographical material on Jacob
        Leupold.

672.   Rose, Paul Lawrence. "Galileo's Theory of Bal-
        listics." **British Journal for the History of**
        Science 4 (1968): 156-59.

        Analyzes Galileo's application of mechanics
        to the theory of ballistics.

673.   _____, and Stillman Drake. "The Pseudo-
        Aristotelian **Questions of Mechanics** in Re-
        naissance Culture." **Studies in the Renais-**
        sance 18 (1971): 65-104.

        Traces the bibliographic background of the
        Mechanica. Includes a section analyzing the in-
        fluence of the **Mechanica** on Renaissance engi-
        neers.

674.   Rossi, Paolo. **Philosophy, Technology, and the**
        Arts in the Early Modern Era. Translated by
        Salvator Attansio. Edited by Benjamin
        Nelson. New York: Harper & Row, 1970.

        Focuses on the role of Francis Bacon and the
        influence of his philosophy of science on the de-
        velopment of technology. Emphasizes the insepa-
        rability of truth and utility.

675.   Scaglia, Gustina, ed. **Taccola's De machinis.**
       The Engineering Treatise of 1449. 2 Volumes.
       Wiesbaden: Reichert, 1971.

       Provides a facsimile edition of 15th century
       book on machines. Also includes an introduction,
       description of engines and technical commentaries
       by the editor.

676.   Schimank, Hans, ed. and trans. **Otto von**
       Guerickes neue (sogenannte) Magdeburger
       Versuche uber den leeren Raum. Dusseldorf:
       VDI Verlag, 1968.

       Provides a German translation of von
       Guericke's treatise on the vacuum along with
       source material and commentary. Shows the rela-
       tionship between von Guericke's natural philoso-
       phy and his inventions. Only translation into a
       modern language that is available.

677.   Shelby, Lon R. **Gothic Design Techniques: The**
       Fifteenth-Century Design Booklets of Mathes
       Roriczer and Hanns Schmuttermayer.
       Carbondale: Southern Illinois University
       Press, 1977.

       Analyzes the designing of Gothic cathedrals
       by the art of geometry. Argues that masons used
       a constructive geometry in which physical tem-
       plates, compasses, and straight-edges were used
       to manipulate geometrical figures. Argues that
       design booklets using such geometry were the be-
       ginnings of a theory of architectural practice.

678.   Stiegler, Karl. "Einige Probleme der Elastiz-
       itatstheorie im 17. Jahrhundert." **Janus** 56
       (1969): 107-122.

       Describes the application of mathematics to
       problems of elasticity during the 17th century.

679.   Stroup, Alice. "Christiaan Huygens and the De-
       velopment of the Air Pump." **Janus** 68
       (1981): 129-58.

       Discusses the contribution of Huygens' scien-
       tific theories to the development of the air
       pump.

680. Talbot, G.R., and A.J. Pacey. "Antecedents of
      Thermodynamics in the Work of Guillaume
      Amontons." **Centaurus** 16 (1971): 20-40.

      Discusses the work of Guillaume Amontons
      (1663-1705) on temperature, gases, and heat en-
      gines.

681. Thorndike, Lynn. "Marianus Jacobus Taccola."
      **Archives Internationales d'Histoire des Sci-
      ences** 8 (1955): 7-26.

      Describes Taccola's contribution to the study
      of machines in the 15th century.

682. Westfall, Richard S. "Robert Hooke, Mechanical
      Technology and Scientific Investigation."
      The Uses of Science in the Age of Newton
      (item 586), pp. 85-110.

      Discusses the role of mechanical technology
      in the work of Robert Hooke. Argues that Hooke
      was a Baconian who believed that science should
      satisfy human needs but that immediate applica-
      tion of science might restrict its progress.
      Concludes that Hooke's contributions to technol-
      ogy did not arise from the application of sci-
      ence.

683. White, Lynn, Jr.. "Kyeser's 'Bellifortis': The
      First Technological Treatise of the Fifteenth
      Century." **Technology and Culture** 10 (1969):
      436-41.

      Discusses the publication of Kyeser's
      Bellifortis (see item 639). Shows that Kyeser
      saw magical knowledge as a branch of technology.

684. Williams, E. "Some Observations of Leonardo,
      Galileo, Mariotte and Others Relative to Size
      Effect." **Annals of Science** 13 (1957): 23-
      29.

      Describes early work on the scale effect.

See also: 13, 15, 22, 24, 34, 37, 48, 64, 70, 76, 79,
160, 163, 817, 1051, 1334, 1432, 1434.

CHAPTER VI: APPLIED MECHANICS

GENERAL WORKS

Primary Sources

685.   Alembert, Jean Lerond, d'.   **Traité de dynamique.**
          Paris: David, 1743.

          Fundamental work on the theory of dynamics.
       Influenced the development of engineering sci-
       ence.

686.   Armengaud, Jacques Eugene.   **The Practical
          Draughtsman's Book of Industrial Design, and
          Machinist's and Engineer's Drawing Companion:
          Forming a Complete Course of Mechanical, En-
          gineering, and Architectural Drawing.**   Trans-
       lated and rewritten by William Johnson.   New
       York: Stringer & Townsend, c. 1854.

          American edition of a European work on engi-
       neering drawing.   Helped to raise American draw-
       ing to the level of European.

687.   Atwood, George.   **A Treatise on the Rectilinear
          Motion and Rotation of Bodies; With a De-
          scription of Original Experiments Relative to
          the Subject.**   Cambridge: J. & J. Merill,
       1784.

          Early theoretical work on mechanics.

688.   Barlow, Peter.   **The Encyclopaedia of Arts, Manu-
          factures, and Machinery.**   London: J.J.
       Griffin, 1848.

          Summary of applied mechanics.

689.   _____. **The Encyclopaedia of Mechanical Philoso-
        phy: Comprising the Sciences of Mechanics,
        Hydrodynamics, Pneumatics and Optics.**
        London: Griffin, 1848.

        Survey of applied mechanics. Based on his
        essay in the **Encyclopaedia Metropolitana** (item
        690).

690.   _____. "Mechanics." **Encyclopaedia Metropoli-
        tana.** Volume III, 1845.

        Provides an overview of applied mechanics
        during the 19th century. Describes work done on
        the strength of materials at the Royal Dockyard
        at Woolwich. Also describes tests done by Thomas
        Telford on the strength of iron using a chain-
        testing machine.

691.   Becker, Max. **Handbuch der Ingenieur-
        Wissenschaft.** 5 vols. Stuttgart: C. Macken,
        1856-61.

        One of the earliest handbooks on engineering
        science. Includes, Volume 1: **Allgemeine Baukunde
        des Ingenieurs**, Volume 2: **Der Bruckenbau**, Volume
        3: **Der Strassen- und Eisenbahnbau**, Volume 4: **Der
        Wasserbau**, Volume 5: **Ausgefuhrte constructionen
        des Ingenieurs.** Went through several later edi-
        tions. Also includes five volumes of plates.

692.   Beckmann, Johann. **Anleitung zür Technologie,
        oder zür Kentniss der Handwerke, Fabriken und
        Manufacturen.** Gottingen, 1777.

        One of the earliest books to treat technology
        as a university subject.

693.   Belanger, Jean Baptiste Charles Joseph. **Cours de
        mécanique, ou resume de leçons sur la dy-
        namique, la statique, et leurs applications à
        l'art de l'ingénieur.** Paris: C. Goeury & V.
        Dalmont, 1847.

        Course on applied mechanics by an important
        French engineering scientist.

694.   Belidor, Bernard Forest de. **Nouveau cours de
        mathématique, à l'usage de l'artillerie et du
        genie où l'on applique les parties les plus
        utiles de cette science à la théorie & à la**

pratique des différents sujets que peuvent
avoir rapport à la guerre. Paris: C.A.
Jombert, 1725.

Important 18th-century work on the application of the science of mechanics to ballistics.

695.   Bernoulli, Christoph. **Cours de mécanique pratique à l'usage des directeurs et con-
tremaîtres de fabrique.** Brusselles: Journal
des Economistes, 1846.

Course on applied mechanics.

696.   _____. **Handbuch der Technologie.** Basel:
Schweighauser, 1833.

Handbook on technology.

697.   Biot, Jean Baptiste. **An Elementary Treatise on
Analytical Geometry, Translated from the
French of J.B. Biot, for the Use of the
Cadets of the Virginia Military Institute,** at
**Lexington, Va.; and Adapted to the Present
State of Mathematical Instruction in the Col-
leges of the United States.** Translated by
Francis H. Smith. New York: Wiley and
Putnam, 1840.

American edition of an important French work
on analytical geometry.

698.   _____. **Notions élémentaires de statique desti-
nées aux jeunes gens qui se preparent pour
l'École polytechnique, ou qui suivent les
cours de l'École militaire de St.Cyr.** Paris:
Bachelier, 1829.

Presents an elementary theory of statics for
students.

699.   Borgnis, Giuseppe Antonio (Joseph A.). **Dictionnaire de mécanique appliquée aux arts: Contenant la definition et la description som-
maire des objets les plus importants ou les
plus usités qui se rapportent à cette science, l'énoncé de leurs propriétés essen-
tielles, et des indications qui facilitent la
recherche des details plus circonstanciés;
ouvrage faisant suite au traité complet de**

mécanique appliquée aux arts. Paris:
Bachelier, 1823.

Dictionary of mechanical engineering and
applied mechanics.

700. _____. Théorie de la mécanique usuelle, ou Introduction à l'étude de la mécanique appliquée aux arts; contenant les principes de
statique, de dynamiquée, d'hydrostatique et
d'hydrodynamique, applicables aux arts indus-
triels; la théorie des moteurs, des effets
utiles des machines, des organes mécaniques
intermédiaires, et l'équilibre des supports.
Paris: Bachelier, 1821.

Important work on applied mechanics and the
theory of machines. Particularly strong on hydrostatics and hydrodynamics.

701. _____. Traité complet de mécanique appliquée
aux arts, contenant l'exposition méthodique
des théories et des expériences les plus
utiles pour diriger le choix, l'invention, la
construction et l'emploi de toutes les es-
pèces de machines. 8 volumes. Paris:
Bachelier, 1818-20.

Detailed study of mechanical engineering.
Includes material on the theory of machines, hydraulic machines and construction.

702. Bossut, Charles. Cours de mathématiques à
l'usage des Écoles royales militaires;
mécanique et hydrodynamique. Paris: Jombert,
1782.

Textbook used at Mézières on the application
of mathematics to mechanics and hydrodynamics.

703. _____. Traité élémentaire de mécanique. Paris:
C.-A. Jombert, 1775.

Work on applied mechanics by the professor of
mathematics at the École Royale Militaire at
Mézières.

704. Burns, Robert Scott. Mechanics and Mechanism:
Being Elementary Essays and Examples for the
Use of Schools, Students, and Artisans.
London: Ingram, Cooke, and Co., 1853.

Useful introduction to the role of mechanics
in the mid-19th century. Went through seven editions.

705.   Camus, Charles Étienne Louis. **Éléments
       d'arithmétique.** Paris, 1749.

       Part of a mathematics course for engineering
       students at the École du Génie at Mézières.

706.   _____. **Élements de géométrie théorique et
       pratique.** Paris, 1750.

       Part of a mathematics course for engineering
       students at the École du Génie at Mézières.

707.   _____. **Traité des forces mouvantes, pour la
       practique des arts et métiers, avec une
       explication de vingt machines nouvelles &
       utiles.** Paris: C. Jombert et L. LeConte,
       1722.

       Early work on applied mechanics for engi-
       neers.

708.   Carnot, Lazare Nicolas Marguerite. **Géométrie de
       position.** Paris: J.B.M. Duprat, 1803.

       Early work on projective geometry by a great
       engineering scientist and father of Sadi Carnot.

709.   Carpenter, Rolla C. **Experimental Engineering:
       For Engineers and for Students in Engineering
       Laboratories.** 2nd revised edition. New
       York: John Wiley & Sons, 1893.

       First single volume work to contain the stan-
       dard methods for testing materials, engines and
       machinery. Written by a Cornell engineering pro-
       fessor. First published in 1890 as **Notes to Me-
       chanical Laboratory Practice.**

710.   Cauchy, Augustin Louis. **Cours d'analyse et
       mécanique de l'École polytechnique.** Paris:
       Imprimerie royale, 1827.

       Lectures on analysis and mechanics given at
       the École Polytechnique.

711.   _____. "Démonstration analytique d'une loi
          découverte par M. Savart et relative aux
          vibrations des corps solides ou fluides."
          Mémoires de l'Académie Royale des Sciences
          2nd ser. 9 (1830): 117-118.

          Puts forward an important theory of vibra-
          tion.

712.   _____. "Mémoire sur la torsion et les
          vibrations tourantes d'une verge retangu-
          laire." **Mémoires de l'Académie Royale des
          Sciences** 2nd ser. 9 (1830): 119-24.

          Important paper on the theory of torsion and
          vibration.

713.   Christian, Gérard Joseph. **Traité de mécanique
          industrielle, ou Exposé de la science de la
          mécanique déduite de l'expérience et de
          l'observation; principalement à l'usage des
          manufacturiers et des artistes.** 3 volumes.
          Paris:Bachelier, 1822-25.

          Important work on the role of applied mechan-
          ics in mechanical engineering. Also includes a
          discussion of the steam engine.

714.   Church, Irving P. **Mechanics of Engineering.
          Comprising Statics and Dynamics of Solids;
          the Mechanics of the Materials of Construc-
          tion, or Strength and Elasticity of Beams,
          Columns, Shafts, Arches, etc.; and the Prin-
          ciples of Hydraulics and Pneumatics, with Ap-
          plications. For Use in Technical Schools.**
          New York: John Wiley & Sons, 1894.

          Combines the author's previous works on stat-
          ics, dynamics, mechanics of materials and mechan-
          ics of fluids into one text. Author was a pro-
          fessor of applied mechanics at the College of En-
          gineering, Cornell University.

715.   Clebsch, Alfred. **Analytische Mechanik.** 2 vol-
          umes. Karlsruhe, 1859.

          Based on lectures at the Karlsruhe Polytech-
          nic by the professor of theoretical mechanics and
          a leading engineering scientist.

716. Cotterill, James H. **Applied Mechanics: An Ele-
      mentary General Introduction to the Theory** of
      Structures. London: Macmillan & Co., 1884.

      Widely used textbook on applied mechanics.
      Contains sections on mechanical engineering, ma-
      chinery, kinematics, gearing, strength of materi-
      als, hydraulics and compressed air. Went through
      several editions.

717. Davies, Charles. **Elements of Surveying, and Nav-
      igation, with Descriptions of the Instruments
      and the Necessary Tables.** Revised edition.
      New York: A.S. Barnes & Co., 1855.

      Advocates the use of sign convention adopted
      by the U.S. Topographical Bureau. Written for
      use of students at West Point.

718. Delius, Christoph Traugott. **Anleitung zu der
      bergbaukunst nach ihrer Theorie und ausübung,
      nebst einer abhandlung von den grundsätzen
      der bergwerks-kammeralwissenschaft, fur die
      Kairerl. Königh Schemnitzer Bergakademie.**
      Vienna: J.T. Edlen, 1773.

      Based on lectures at the Mining School at
      Schemnitz, Hungary, one of the first modern tech-
      nical schools. Its syllabus influenced the École
      Polytechnique. Includes a discussion of air and
      steam engines. Translated into French in 1778.

719. Desaguliers, John Theophilus. **A Course of Exper-
      imental Philosophy.** 2 volumes. London: J.
      Senex, 1734-44.

      Early work on natural philosophy with appli-
      cations to engineering Work influenced many
      early engineering scientists.

720. _____. **Lectures of Experimental Philosophy,
      wherein the Principles of Mechanicks, Hydro-
      staticks, and Optics, are Demonstrated and
      Explained at Large, by a Great Number of Cu-
      rious Experiments.** London: W. Mears, B.
      Creake, and J. Sackfield, 1719.

      Early work on experimental natural philosophy
      with applications to engineering.

721. Deschanel, Augustin Privat. **Elementary Treatise**
      on Natural Philosophy. 4 volumes. Trans-
      lated and edited by J.D. Everett. New York:
      D. Appleton & Co., 1881-1883.

      English translation of a French work pub-
      lished in 1868. Influenced many university-
      trained engineers.

722. Duhamel, Jean Marie Constant. **Cours d'analyse de**
      l'École polytechnique. Paris: Bachelier,
      1840.

      Course on mathematics taught for engineers at
      the École Polytechnique.

723. _____. **Cours de mécanique.** 2 volumes. Paris:
      Bachelier, 1845-46.

      Important work on applied mechanics based on
      lectures given at the École Polytechnique. Went
      through numerous editions. Translated into
      German in 1843.

724. Eddy, Henry Turner. **Researches in Graphical**
      Statics. New York, 1878.

      Presents an extension of graphical methods to
      the area of statics. One of the first American
      works on the subject, and one of the first Ameri-
      can engineering books to be translated into
      German.

725. Euler, Leonhard. **Theoria motvs corporvm soli-**
      dorvm sev rigidorvm ex primis mostrae cogni-
      tionis principiis stabilita et ad omnes
      motvs, qvi hvivsmodi corpora cadere possvnt
      accommodata. Rostochii: A.F. Rose, 1765.

      Fundamental work in Latin on the principle of
      stability and the theory of dynamics.

726. Eytelwein, Johann Albert. **Aufgaben, grössten-**
      theils aus der angewandten Mathematik zür Ue-
      bung der Analysis. Für angehende Feldmesser,
      Ingenieurs und Baumeister. Berlin: Friedrich
      Maurer, 1793.

      Early work on mechanics for engineers by the
      director of the Berlin Bauakademie.

727.   _____. **Handbuch der Mechanik fester Körper und der Hydraulik.** Berlin: F.T. Lagarde, 1801.

Important German contribution to applied mechanics and hydraulics by the director of the Berlin Bauakademie.

728.   _____. **Handbuch der Statik fester Körper.** Berlin: Realschulbuchhandlung, 1808.

Important German work on engineering statics by the director of the Berlin Bauakademie.

729.   Fairbairn, William. **Useful Information for Engineers.** London: Longmans, Brown, Green and Longmans, 1856.

Series of lectures on the applied sciences. Also includes a treatise on iron roofs, lectures on the machinery of the Paris Universal Exhibition of 1855, and on the effects of impacts and vibrations on girders.

730.   Farrar, John. **An Elementary Treatise on Mechanics, Comprehending the Doctrine of Equilibrium and Motion, as Applied to Solids and Fluids, Chiefly Compiled, and Designed for the Use of Students for the University at Cambridge, New England.** 2nd edition. Boston: Hilliard, Gray, and Co., 1834.

Early Harvard textbook on applied mechanics. First edition published in 1825.

731.   Faunce, Linus. **Mechanical Drawing. Prepared for the Use of the Students of the Massachusetts Institute of Technology.** Boston: published by the author, 1887.

Early textbook from M.I.T. Went through 11 editions.

732.   Föppl, August. **Vorlesungen uber technische Mechanik.** 6 volumes. Leipzig: B.G. Teubner, 1897-1910.

Important work on applied mechanics by a leading engineering scientist and teacher of S. Timoshenko and L. Prandtl. Based on his lectures as professor of mechanics at the Polytechnical Institute of Munich. Volume I: **Introduction to**

Mechanics; Volume II: **Graphical Statics;** Volume
III: **Mechanics of Materials;** Volume IV: **Dynamics;**
Volume V: **The Most Important Studies of Advanced
Elasticity Theory;** Volume VI: **The Most Important
Studies of Advanced Dynamics.**

733. Gerstner, Franz Joseph. **Handbuch der Mechanik.**
      3 volumes. Prague: J. Spurny, 1832-34.

      Important book on applied mechanics by the
      first director of the Bomische Institute.

734. Gillespie, William Mitchell. **A Manual of the
      Principles and Practice of Road-Making: Com-
      prising the Location, Construction, and Im-
      provement of Roads, (Common, Macadam, Paved,
      Plank, etc.) and Rail-Roads.** New York: A.S.
      Barnes & Co., 1847.

      Important scholarly work on the application
      of mechanics to road building by a professor at
      Union College.

735. Goodeve, T.M. **Principles of Mechanics.** New
      York: D. Appleton and Co., 1876.

      Science textbook adapted for the use of
      artisans.

736. Gordon, Lewis D.B. **A Syllabus of a Course of
      Lectures on Civil Engineering and Mechanics.**
      Edinburgh: T. Constable, 1841.

      Syllabus for a course on civil engineering
      and mechanics at the University of Glasgow by the
      first engineering professor in Great Britain.

737. _____. **A Synopsis of Lectures on Civil Engi-
      neering and Mechanics.** Glasgow: R. Griffin &
      Co., 1849.

      Synopsis of lectures on civil engineering and
      mechanics at the University of Glasgow.

738. Hutton Charles. **A Course of Mathematics in Two
      Volumes, Composed and More Especially De-
      signed for Use of the Gentlemen Cadets in the
      Royal Military Academy at Woolwich.** London:
      G.G. and J. Robinson, 1798-1801.

Mathematical textbook used at Woolwich. Included practical applications. Went through numerous editions.

739. Jackson, Thomas. **Elements of Theoretical Mechan-
      ics; Being the Substance of a Course of Lec-
      tures on Statics and Dynamics.** Edinburgh: W.
      & D. Laing, 1827.

      Early 19th-century work on mechanics.

740. Jamieson, Alexander. **A Dictionary of Mechanical
      Science, Arts, Manufactures, and Miscella-
      neous Knowledge.** 2 volumes. London: H.
      Fisher, Son and Co., 1827.

741. Kelvin, William Thomson, Baron, and Peter
      Gutherie Tait. **Elements of Natural Philoso-
      phy.** Cambridge: Cambridge University Press,
      1873.

      Influential textbook on analytic mechanics.
      Went through several editions.

742. Lagrange, Joseph Louis. **Mécanique analytique.**
      Paris: Chez la Veuve Desaint, 1788.

      Important and fundamental work on mechanics.
      Went through several editions and was translated
      into German, Russian and English.

743. La Hire, P. de. **Traité de mécanique.** Paris,
      1695.

      Early work on applied mechanics. Uses geo-
      metric methods for mechanical analysis. Provided
      an approach to the analysis of structures.

744. Lamb, Horace. **Statics, Including Hydrostatics
      and the Elements of the Theory of Elasticity.**
      Cambridge: Cambridge University Press, 1912.

      Influential textbook on the mathematical the-
      ory of hydrostatics and the theory of elasticity.

745. Lamé, Gabriel. **Cours de physique de l'École
      polytechnique.** 2 volumes. Paris: Bachelier,
      1836-37.

      Course on physics taught at the École Poly-
      technique. Includes material on the properties

of bodies, the theory of heat, acoustics, the theory of light and electricity and magnetism. Went through several editions. Influenced engineering science. Translated into German.

746.   Lanza, Gaetano. **Applied Mechanics**. New York: John Wiley & Sons, 1885.

Textbook on the strength of materials and the stability of structures used at M.I.T.

747.   Lardner, Dionysius. **Hand-Book of Natural Philosophy and Astronomy**. 3 volumes. Philadelphia: Blanchard & Lea, 1851-54.

Popular books which presented scientific ideas and influenced engineers and mechanics. Includes sections on mechanics, hydrostatics, hydraulics, and heat. Went through several editions and some parts, such as those on heat and hydrostatics, were published separately.

748.   _____, and Henry Kater. **A Treatise on Mechanics**. London: Longman, Brown, Green & Longman, [c.1844].

Scientific treatise on mechanics that influenced many engineers.

749.   McAdam, John Loudon. **A Practical Essay on the Scientific Repair and Preservation of Public Roads**. London: B. Macmillan, 1819

Important early work on the application of science to road building.

750.   Mahan, Dennis H. **Descriptive Geometry, as Applied to the Drawing of Fortification and Stereotomy**. New York: J. Wiley, 1864.

Work on descriptive geometry used as a textbook at West Point. Went through several editions.

751.   _____. **Industrial Drawing: Comprising the Description and Uses of Drawing Instruments, the Construction of Plane Figures, Tinting, the Projections and Sections of Geometrical Solids, Shadows, Shading, Isometrical Drawing, Oblique Projection, Perspective, Architectural Elements, Mechanical and Topographi-**

cal Drawing. New York: John Wiley & Sons, 1852.

Work on descriptive geometry by an important professor of military and civil engineering at West Point.

752.   Maxwell, James Clerk. **Matter and Motion.** London: Society for Promoting Christian Knowledge, 1876.

Important book on the role of force and energy in mechanics. Went through several editions and influenced engineering science. Discusses the kinetic theory of heat.

753.   _____. "On a Dynamical Top, for Exhibiting the Phenomena of the Motion of a System of Invariable Form About a Fixed Point, with Some Suggestions as to the Earth's Motion." **Transactions of the Royal Society of Edinburgh** 21 (1857): 559-70.

Important analysis of the dynamics of circular motion.

754.   Mohr, Otto Christian. **Abhandlungen aus dem Gebiete der technischen Mechanik.** Berlin: W. Ernst, 1906.

Important work on engineering mechanics by the professor of railway structures, hydraulics and earth structures at the Dresden Polytechnik. Includes important work on the theory of elasticity.

755.   _____. **Technische mechanik.** Stuttgart: K. Wittwer, 1878.

Based on lectures the author gave while holding the chair in engineering mechanics at the Stuttgart Polytechnic. Includes important work on the theory of elasticity.

756.   Monge, Gaspard. **Application de l'analyse à la géométrie, à l'usage de l'école impériale polytechnique.** Paris: Bernard, 1807.

Important work on the application of descriptive geometry to engineering. Used at the École Polytechnique.

757.   _____. **Géométrie descriptive.** Paris: Baudouin,
        1799.

        Fundamental work on descriptive geometry.
        Influential at the École Polytechnique.

758.   Morin, Arthur Jules. **Aide-mémoire de mécanique
        pratique à l'usage des officiers d'artillerie
        et des ingénieurs civil et militaires.** Metz:
        Theil, 1837.

        Handbook on mechanics.

759.   _____. **Leçons de mécanique pratique.** Paris: L.
        Hachette, 1853.

        Eventually contained five volumes including
        Volume 1: **Notions fondamentales de mécanique et
        données d'expérience.** Volume 2: **Hydraulique.**
        Volume 3: **Des machines à vapeur.** Volume 4: **Ré-
        sistance des materiaux.** Volume 5: **Notions
        géométriques sur les mouvements et leurs
        transformations, ou elements de cinématique.**
        Important work in French engineering science.
        Applies the concept of <u>vis viva</u> to machines.
        American edition translated by Joseph Bennett
        under the title **Fundamental Ideas of Mechanics
        and Experimental Data.** New York: D. Appleton,
        1860.   Influenced the work of James B. Francis.

760.   _____. **Nouvelles expériences sur le frottement,
        faites à Metz en 1831-1833.** 3 volumes.
        Paris: Bachelier, 1832-35.

        Experimental studies on engineering science
        conducted at Metz.   Includes a study of friction
        and a study of the impact of projectiles.

761.   Moseley, Henry. **Illustrations of Mechanics.** Re-
        vised by James Renwick. New York: Harper,
        1839.

        Early work on mechanics by an important fig-
        ure in the development of a theoretical approach
        to technology.

762.   _____. **The Mechanical Principles of Engineering
        and Architecture.** London: Longman, Brown,
        Green and Longmans, 1843.

Important work on applied mechanics. One of
the first textbooks to apply the concept of work
to machines. Based on lectures given at King's
College in 1840-42. American edition with additions by D.H. Mahan published in New York by
Wiley & Halsted in 1856.

763. _____. A Treatise on Mechanics, Applied to the
      Arts; Including Statics and Hydrostatics.
      London: J.W. Parker, 1834.

      Early textbook on applied mechanics.

764. Moxon, Joseph. Mechanick Exercises. Introduc-
      tion by Benno M. Forman. Edited by Charles
      F. Montgomery. New York: Praeger, 1970.

      Reprint of the 1703 edition. Early work on
      applied mechanics.

765. Navier, Claude Louis Marie Henri. Résumé des
      leçons d'analyse données à l'École polytech-
      nique. Paris: Carilian-Goeury et V. Dalmont,
      1840.

      Presents a course on calculus taught to engi-
      neers to the École Polytechnique.

766. _____. Résumé des leçons de mécanique données à
      l'École polytechnique. Paris: Carilian-
      Goeury et V. Dalmont, 1841.

      Presents a survey of a course on applied me-
      chanics given at the École Polytechnique. Later
      translated into German.

767. _____. Résumé des leçons données à l'école des
      ponts et chaussées sur l'application de la
      mecanique à l'établissement des constructions
      et machines. 2nd edition. Paris: Carilian-
      Goeury, 1833.

      Presents a survey of applied mechanics by one
      of the leading engineering scientists in France.
      Applies advanced mathematics to technical prob-
      lems.

768. Poisson, Siméon Denis. "Mémoire sur le mouvement
      d'un corps solide." Mémoires de l'Académie
      Royale des Sciences 2nd ser. 14 (1838): 275-
      432.

Significant contribution to dynamics by an
important theoretician of engineering.

769.   _____. **Traité de mécanique.** Paris: Courcier,
        1811.

Important work on mechanics. Used as a textbook at the École Polytechnique. Went through
several editions and was published in English as
A Treatise of Mechanics. London: Longman & Co.,
1842. Also translated into German.

770.   Poncelet, Jean Victor. **Cours de mécanique indus-
        trielle, professé de 1828 à 1829.** Metz,
        1829.

Set of lectures on applied mechanics. Deals
with both statics and dynamics. Many later editions published.

771.   _____. **Traité des propriétés projectives fig-
        ures; ouvrage utile à ceux qui s'occupent des
        applications de la géométrie descriptive et
        d'operations géométriques sur la terrain.**
        Paris: Bachelier, 1822.

Work on descriptive geometry by a leading
French engineering scientist.

772.   Prony, Gaspard Clair François Marie Riche, Baron
        de. **Leçons de mécanique analytique données à
        l'École impériale polytechnique.** 2 volumes.
        Paris: Impr. de l'École impériale des ponts
        et chaussées, 1810-1815.

Important work on analytic mechanics by a
member of the École Polytechnique.

773.   _____. **Mécanique philosophique, ou analyse
        raisonnée des diverses parties de la science
        de l'équilibre et du mouvement.** Paris: Im-
        primerie de la République, 1800.

Fundamental work an mechanics which influenced engineering science.

774.   _____. **Sommaires de leçons sur le mouvement des
        corps solides, l'équilibre, et le mouvement
        des fluides, données à l'École impériale
        polytechnique en 1809.** Paris, 1809.

Lectures by a leading French engineering scientist on the dynamics of solids and fluids.

775. Rankine, William John Macquorn. **A Manual of Applied Mechanics.** London: Richard Griffin & Co., 1858.

Important and exhaustive survey of applied mechanics. Presents some of the first studies of stress and strain to English-speaking engineers. Used as a textbook in Europe and America well into the 20th century. Went through several editions and was translated into several languages.

776. _____. "On the Stability of Loose Earth." **Philosophical Transactions of the Royal Society of London** 147 (1857): 9-28.

Important paper on the application of applied mechanics to the theory of the stability of earthworks.

777. Renwick, James. **Applications of the Science of Mechanics to Practical Purposes.** New York, 1842.

Deals with machinery, railroads, canals, wheels, roads, ships, and mining. Provides principles of design. Applies the methods of science to technology but makes very little use of Renwick's earlier work (see item 778).

778. _____. **The Elements of Mechanics.** Philadelphia, 1832.

Attempts to bridge the gap between science and technology. Uses Newtonian mechanics. Argues for the application of the contents of science to technology.

779. Reynolds, Osborne. **Syllabus of the Lectures in Engineering at the Owens College.** Manchester: J.E. Cornish, 1875.

Series of lectures by a leading engineering scientist who made major contributions to fluid dynamics and aerodynamics.

780. Robison, John. **A System of Mechanical Philoso-
      phy, with Notes by David Brewster.** 4 vol-
      umes. Edinburgh: J. Murray, 1822.

      Influential work on the mechanical philoso-
      phy, containing applications to engineering. In-
      fluenced many early engineering scientists, par-
      ticularly in the area of the application of sci-
      ence to the steam engine. Contains works on me-
      chanics and on the steam engine, based on arti-
      cles in the **Encyclopaedia Britannica.** Includes
      notes by James Watt.

781. Saint-Venant, Adhémar Jean Claude Barré de.
      Leçons de mécanique appliqué. Paris, 1838.

      Work an applied mechanics based on a course
      at the École des Ponts et Chaussées.

782. _____. **Principes de mécanique fondés sur la
      cinématique.** Paris: Bachelier, 1851.

      Based on lectures on applied mechanics given
      by a leading figure in the theory of elasticity.

783. Terzaghi, Karl. **Erdbaumechanik auf boden-
      physikalischer grundlage.** Leipzig: F.
      Deuticke, 1925.

      Fundamental work on the engineering science
      of soil mechanics.

784. _____. **Principles of Soil Mechanics.** New York:
      McGraw-Hill, 1926.

      Based on work in the **Engineering News-Record**
      for 1925.

785. _____. **Theoretical Soil Mechanics.** New York:
      J. Wiley & Sons, 1943.

      Fundamental textbook on the engineering sci-
      ence of soil mechanics.

786. Ure, Andrew. **A Dictionary of Arts, Manufactures,
      and Mines: Containing a Clear Exposition of
      Their Principles and Practice.** London:
      Longman, Orme, Brown, Green & Longmans, 1839.

      Important compendium of the application of
      science to the practical arts.

787. Von Karman, Theodore, and Maurice Biot. **Mathe-
      matical Methods in Engineering: An Introduc-
      tion to the Mathematical Treatment of Engi-
      neering Problems.** New York: McGraw-Hill,
      1940.

          Important work on the application of differ-
      ential equations, Fourier's series, and opera-
      tional calculus to engineering problems by a ma-
      jor contributor to the theory of aerodynamics.
      Translated into French and Spanish.

788. Warren, S. Edward. **A Manual of Elementary Pro-
      jection Drawing, Involving Three Dimensions.**
      4th edition, revised. New York: John Wiley &
      Sons, 1873.

          Textbook for beginning students by a profes-
      sor at M.I.T. First published in 1861 as **A Man-
      ual for Elementary Geometrical Drawing.**

789. Weisbach, Julius. **Der Ingenieur. Sammlung von
      Tafeln, Fromeln und Regeln der Arithmetik,
      Geometrie und Mechanik.** Braunschweig: F.
      Vieweg, 1848.

          Handbook on engineering and applied mechan-
      ics. Went through numerous editions, with later
      editions edited by Franz Reuleaux.

790. _____. **Lehrbuch der Ingenieur- und Machinen-
      Mechanik.** 3 volumes. Braunschweig: F.
      Vieweg und Sohn, 1850-60.

          Important textbook on applied mechanics and
      mechanical engineering. Contains work on theo-
      retical mechanics, statics, the theory of ma-
      chines, the steam engine, and hydraulics. Based
      on the calculus. Went through numerous editions.
      Translated into English as A Manual of the **Me-
      chanics of Engineering and of the Construction of
      Machines.** New York: D. Van Nostrand, 1870. Used
      as a textbook in American engineering schools.
      Volume 1 was translated into English as **Mechanics
      of Engineering: Theoretical Mechanics, with an
      Introduction to the Calculus.** English transla-
      tions went through several editions.

791.   _____. **Principles of the Mechanics of Machinery
        and Engineering.** 2 volumes. London: H.
        Bailliere, 1847-48.

        English translation of lectures on mechanics
        given by Weisbach. Volume 1: **Theoretical Mechan-
        ics.** Volume 2: **The Application of Mechanics to
        Machinery.** Volume 2 was translated by Lewis
        Gordon, first Regius Professor of Civil Engineer-
        ing and Mechanics at the University of Glasgow.
        Also published in an American edition in
        Philadelphia by Lea & Blanchard, 1848-49.

792.   Whewell, William. **Analytic Statics, A Supplement
        to the Fourth Edition of An Elementary Trea-
        tise on Mechanics.** Cambridge: J. & J.J.
        Deighton, 1833.

        Presents a theory of statics using higher
        mathematics and virtual velocities. Applies the
        theory to some engineering problems, such as
        those involving the suspension bridge.

793.   _____. **An Elementary Treatise on Mechanics: De-
        signed for the Use of Students in the Univer-
        sity.** 5th edition. Cambridge: J. & J.J.
        Deighton, 1836.

        Includes a theory of arches and a theory of
        work in the steam engine. Employs graphical so-
        lutions but no calculus. Used as a textbook at
        Cambridge.

794.   _____. **The First Principles of Mechanics, with
        Historical and Practical Illustrations.**
        Cambridge: J. & J.J. Deighton, 1832.

        A popular treatise in which the theory of me-
        chanics is developed and applied to practical
        problems. Uses only geometrical methods.

795.   _____. **The Mechanics of Engineering, Intended
        for Use in Universities and in Colleges of
        Engineers.** Cambridge: J. & J. Deighton,
        1841.

        Early engineering work by a leading philoso-
        pher of science.

796.   _____. **On the Motion of Points Constrained and Resisted, and on the Motion of a Rigid Body. The Second Part of a New Edition of A Treatise on Dynamics.** Cambridge: J. & J.J. Deighton, 1834.

Develops a theory of dynamics using calculus and geometry.

797.   Wood, De Volson. **The Elements of Analytical Mechanics.** New York: J. Wiley & Sons, 1876.

Widely used American textbook for engineers. Went through many editions.

798.   _____. **The Principles of Elementary Mechanics.** New York: J. Wiley & Sons, 1878.

Popular textbook for engineers.

799.   Young, Thomas. **A Course of Lectures on Natural Philosophy and the Mechanical Arts.** London: J. Johnson, 1807.

Important contribution to applied mechanics. Contains sections on mechanics, hydrodynamics and physics. Modern reprint edition published by Johnson Reprint in 1971.

800.   _____. **A Syllabus of a Course of Lectures on Natural and Experimental Philosophy.** London: Royal Institution, 1802.

Contents of lectures the author gave as professor of natural philosophy at the Royal Institution of Great Britain. Includes sections on applied mechanics.

See also: 389, 394, 407, 621, 639.

Secondary Sources

801.   Artobolevski, I.I. "Evolution de la théorie de la structure et de la classification des mécanismes en Russie depuis le XIXe siècle à nous jours." **Actes VIIIe Congrès International des d'Histoire des Sciences** Paris: Hermann, 1958, pp. 980-83.

Discusses the theory of structures and the classification of mechanisms in Russia during the 19th and 20th centuries.

802.   Bailey, R.W. "The Contributions of Manchester Researches to Mechanical Science." **Institution of Mechanical Engineers** (June 1929): 613-85.

Discusses the contributions to applied mechanics by engineers and scientists from Manchester. Includes a discussion of the role of Eaton Hodgkinson in materials testing.

803.   Booker, Peter Jeffrey. **A History of Engineering Drawing.** London: Chatto & Windus, 1963.

Provides a survey of the history of engineering drawing. Includes a discussion of the contributions of Gaspard Monge and his **Géométrie descriptive** of 1795.

804.   Bradley, Margaret. "Franco-Russian Engineering Links: The Careers of Lame and Clapeyron, 1820-1830." **Annals of Science** 38 (1981): 291-312.

Discusses the contributions of Lame and Clapeyron to Russian engineering during their years of exile. Also discusses their research on the application of science to technology.

805.   Braun, Hans-Joachim. "Franz Reuleaux und der Technologietransfer zwischen Deutschland und Nordamerika am Ausgang des 19. Jahrhunderts." **Technikgeschichte** 48 (1981): 112-130.

Discusses the role of the noted engineering scientist Franz Reuleaux in transferring technology between Germany and North America.

806.   Charbonnier, Prosper. **Essais sur l'histoire de la balistique.** Paris: Société d'éditions géographiques, maritimes et coloniales, 1928.

Provides a history of the application of mechanics to the theory of ballistics.

807. Constant, Edward W., II. "On the Diversity and Co-evolution of Technological Multiples: Steam Turbines and Pelton Water Wheels." **Social Studies of Science** 8 (1978): 183-210.

Investigates the relationship between steam turbines and the Pelton water wheel.

808. Cook, Gilbert. "Rankine and the Theory of Earth Pressure." **Géotechnique** 2 (1950-51): 271-79.

Reviews the contributions of Rankine to the development of a theory of soil mechanics. Concludes that his scientific papers on the subject were highly mathematical and difficult for engineers to understand.

809. Courtel, R., and L.M. Tichvinsky. "On the Development of Theories of Dry and Boundary Friction." **Mechanical Engineering** 85 (September 1963): 55-59, and 85 (October 1963): 33-37.

Surveys theories of friction beginning with 1700.

810. Dugas, René. **A History of Mechanics.** Translated by J.R. Maddox. New York: Central Book Co., 1955.

Useful survey of the history of mechanics. Discusses theories of elasticity, rational mechanics, and fluid mechanics.

811. _____. **La mécanique au 17e siècle.** Neuchâtel: Griffon, 1954.

Survey of 17th-century mechanics. Discusses the work of the Bernoullis, Euler and Lagrange.

812. Duffy, Michael. "Rail Stresses, Impact Loading, and Steam Locomotive Design." **History of Technology** 9 (1984): 43-101.

Discusses the relationship between the study of rail stresses and impact and the design of locomotives.

813. Feldhaus, Franz Maria. **Geschichte des technis-
      chen Zeichnens.** 3rd edition. Edited by
      Edmund Schriff. Wilhelmshaven: Kuhlmann,
      1967.

      Provides a history of mechanical drawing.

814. Fitzgerald, Desmond J. "The Problem of the Pro-
      jectile Again." **Proceedings of the American
      Catholic Philosophical Association** 38
      (1964): 186-201.

815. Glossop, Rudolph. "The Rise of Geotechnology and
      Its Influence on Engineering Practice."
      Géotechnique 18 (1968): 107-150.

      Surveys the history of soil mechanics from
      1700 until the present. Discusses the applica-
      tion of mechanics.

816. Golder, H.Q. "Coulomb and Earth Pressure."
      Géotechnique 1 (1948-49): 66-71.

      Reviews the contributions of Coulomb to the
      theory of soil mechanics. Notes that Coulomb ad-
      vocated neglecting the role of cohesion in calcu-
      lating earth pressures.

817. Hamilton, S.B. "Mechanics in Theory and Prac-
      tice." **Chartered Mechanical Engineer** 14
      (April 1967): 158-64.

      Traces the history of mechanics from the time
      of Aristotle to the late 19th century.

818. Hermann, George, ed. **R.D. Mindlin and Applied
      Mechanics: A Collection of Studies in the
      Development of Applied Mechanics, Dedicated
      to Professor Raymon D. Mindlin by His Former
      Students.** New York: Pergamon, 1974.

      Discusses Mindlin's contributions to applied
      mechanics. Includes historical articles on pho-
      toelasticity and the development of a three di-
      mensional theory of elasticity.

819. Heyman, Jacques. **Coulomb's Memoir on Statics: An
      Essay in the History of Civil Engineering.**
      Cambridge: Cambridge University Press, 1972.

Provides a facsimile English translation and historical analysis of Coulomb's memoir on statics, in which he treated the strength of beams, columns and piers; the theory of earth pressure; and the theory of arches. Analyzes the theories of Coulomb, using both historical and modern engineering techniques. Places the work into a historical context.

820. Institution of Civil Engineers. **A Century of Soil Mechanics: Classic Papers in Soil Mechanics Published by the Institution of Civil Engineers, 1844-1946.** London: Institution of Civil Engineers, 1969.

Provides an insight into a theory of soil mechanics based on applied mechanics.

821. Kerisel, J. "Histoire de la mécanique des sols en France jusqu'au 20e siècle." **Géotechnique** 6 (1956): 151-66.

Surveys the theories of soil mechanics in France during the 17th, 18th, and 19th centuries. Includes an analysis of the contributions of Vauban, Moreau, and Darcy.

822. Klemm, Friedrich. "Die Rolle der Mathematik in der Technik des 19. Jahrhunderts." **Technikgeschichte** 33 (1966): 72-90.

Discusses the role of mathematical analysis in mechanical and civil engineering in 19th-century Germany.

823. Litner, Stephen F., and Darwin H. Stapleton. "Geological Theory and Practice in the Career of Benjamin Henry Latrobe." **Two Hundred Years of Geology in America.** Edited by Cecil J. Schneer. Hanover: University Press of New England, 1979.

Argues that an interest in geology contributed to Latrobe's approach to masonry construction and other engineering problems.

824. Legget, R.F. "'Modern' Studies in Soil Mechanics Go Back Further Than You Think." **Canadian Consulting Engineer** 9 (October 1967): 54-55.

Traces the study of soil mechanics to 1808.

825.   Lorenz, H.   "Die wissenschaftlichen Leistungen F.
       Grashof."   **Beitrage zür Geschichte der Indus-
       trie und Technik**   16 (1926): 1-12.

       Discusses the contributions of Franz Grashof,
       one of the founders of the Verein Deutscher Inge-
       nieure, to applied mechanics.

826.   McKeon, R.   "A Study of the History of 19th-Cen-
       tury Science and Technology: Engineering Sci-
       ence in the Works of Navier."   **Actes du XIIIe
       Congrès International d'Histoire des Sciences**
       11 (1971): 321-27.

       Discusses the interaction of science and
       technology in the works of Navier. Argues that
       his engineering science reflected 19th-century
       attitudes about the role of science in technol-
       ogy.

827.   Mandryka, A.P.   **Istorija ballistiki.**   Moscow:
       Izdatelstvo Akadiemii Nauk SSSR, 1964.

       Traces the history of ballistics to the mid-
       19th century.

828.   Nedoluha, Alois.   **Kulturgeschichte des technis-
       chen Zeichnens.**   Vienna: Springer-Verlag,
       1960.

       Discusses the history of technical drawing
       beginning with the Stone Age. Includes a discus-
       sion of Gaspard Monge's work in the 18th century.
       Includes the names of many 18th- and 19th-century
       contributors to technical drawing.

829.   Skempton, A.W.   "The Publication of Smeaton's Re-
       ports."   **Notes and Records of the Royal Soci-
       ety of London**   26 (1971): 135-55.

       Discusses John Smeaton's reports on civil en-
       gineering, which contained the results of some of
       his engineering experiments.

830.   Taton, René.   "L'École Polytechnique et le nénou-
       veau de la géométrie analytique."   **L'Aventure
       de la Science: Mélanges Alexandre Koyré.**   2
       volumes.   Paris: Hermann, 1964, Volume I, pp.
       552-64.

Discusses the role of the École Polytechnique in the development of analytic geometry.

831. Truesdell, Clifford A. "The Mechanical Foundations of Elasticity and Fluid Dynamics." **Journal of Rational Mechanics** 1 (1952): 125-300.

Survey of the role of rational mechanics in the development of theories of elasticity and fluid dynamics from the time of Hooke to the present. Shows the close connection between elasticity and fluid dynamics. Corrections and additions published in the **Journal of Rational Mechanics** 2 (1953): 593-616.

832. _____. "Reactions of the History of Mechanics upon Modern Research." **Proceedings of the 4th U.S. National Congress of Applied Mechanics.** New York: American Society of Mechanical Engineers, 1962, pp. 35-47.

Includes a discussion of the history of applied mechanics.

833. _____. "Recent Advances in Rational Mechanics." **Science** 127 (1958): 729-39.

Based on the Sigma Xi lecture of 1956. Provides a discussion of theories of elasticity and viscosity. Includes some historical material on Helmholtz, Kelvin, Saint Venant, and others.

See also: 77, 291, 292, 298, 300, 306, 314, 325, 335, 349, 350, 363, 377, 596, 672.

THE STRENGTH OF MATERIALS AND THEORY OF ELASTICITY

Primary Sources

834. Barlow, Peter. **An Essay on the Strength and Stress of Timber, Founded upon Experiments Performed at the Royal Military Academy, on Specimens Selected from the Royal Arsenal, and His Majesty's Dock-yard, Woolwich: Preceded by an Historical Review of Former Theories and Experiments, with Numerous Tables and Plates. Also an Appendix on the Strength**

of Iron, and Other Materials. London: J. Taylor, 1817.

Fundamental work on the experimental study of the strength of materials.

835. _____. **Experiments on the Transverse Strength and Other Properties of Malleable Iron, with Reference to Its Uses for Railway Bars; and a Report Founded on the Same Addressed to the Directors of the London and Birmingham Railway Company.** London: B. Fellowes, 1835.

Experimental study of the strength of iron.

836. _____. "Report on the Present State of Our Knowledge Respecting the Strength of Materials." **British Association for the Advancement of Science** Third Report (1833): 93-103.

Provides an overview of the knowledge of the strength of materials in the early 19th century by one of the leaders in materials testing. Describes tests conducted on timber. Develops a theory of resistance using geometrical arguments.

837. _____. **A Treatise on the Strength of Materials.** Revised by P.W. Barlow and W.H. Barlow. Edited by William Humber. London: Lockwood & Co., 1867.

Fundamental work on the strength of materials. Includes sections on the experimental testing of materials.

838. Bauschinger, Johann. **Elemente der graphischen Statik.** Munich: R. Oldenbourg, 1871.

Work on the application of graphical methods to the theory of elasticity by the professor of mechanics at the Polytechnical Institute of Munich.

839. _____. **Uniform Methods for Testing Construction Materials.** Washington, D.C., 1896.

Work on materials testing by a pioneer in experimental testing of materials.

840.   _____. Versuche uber die festigkeit des
        Bessemer-stahles der walzwerks- und Bessemer-
        stahlfabrikazions-gesellschaft Ternitz,
        angestellt im Mechanisch-technischen labora-
        torium der K. Polytechnischen schule in
        München. Munich: Wolf, 1873.

        Describes the testing of steel in one of the
        first university materials-testing laboratories.

841.   Cauchy, Augustin Louis. "Démonstration analy-
        tique d'une loi découverte par M. Savart et
        relative aux vibrations des corps solides ou
        fluides." Mémoire de l'Académie royale des
        sciences 2nd series 9 (1830): 117-18.

        Contribution to the theory of vibrations of
        solids and liquids by a leading French theoreti-
        cian.

842.   _____. "Mémoires sur la torsion et les vibra-
        tions tournantes d'une verge rectangulaire."
        Mémoires de l'Académie royale des sciences
        2nd series 9 (1830): 119-24.

        Fundamental work on the theory of vibrations
        by a leading French theoretician.

843.   _____. Recherche des équations générales
        d'équilibre pour un système de points ma-
        teriels, assujettis à des liaisons quelcon-
        ques. Paris: Burefrères, 1827.

        Mathematical work on the theory of elasticity
        by a leading French theoretician.

844.   Clebsch, Alfred. Theorie der Elasticität fester
        Körper. Leipzig: B.G. Teubner, 1862.

        Important contribution to the theory of elas-
        ticity by the professor of theoretical mechanics
        at the Karlsruhe Polytechnic. Translated into
        French by A. Saint-Venant.

845.   Coulomb, Charles Augustin de. "Recherches théo-
        rique et expérimentales sur la force de tor-
        sion & sur l'élasticité des fils de metal."
        Mémoires de l'Académie des Sciences (1784):
        229-270.

Paper on the theory of elasticity by a leading French theoretician.

846. Culmann, Karl. **Die graphische Statik.** Zurich:
      Meyer & Zeller, 1866.

      Influential work. Provided engineers with
      graphical methods for analyzing stresses in
      loaded beams.

847. Fairbairn, Sir William. **Iron, Its History, Prop-
      erties, and Processes of Manufacturing.**
      Edinburgh: A. and C. Black, 1861.

      Includes information on the strength of iron
      based on experimental testing.

848. _____. **Treatise on Iron Ship Building: Its
      History and Progress as Comprised in a Series
      of Experimental Researches on the Laws of
      Strain; Including the Experimental Results on
      the Resisting Powers of Armour Plates and
      Shot at High Velocities.** London: Longmans,
      Green & Co., 1865.

      Fundamental work on the strength of iron
      based on experimental research.

849. Flamant, Alfred Aimé. **Stabilité des construc-
      tions, résistance des materiaux.** Paris:
      Baudry, 1886.

      Includes discussion of the theory of struc-
      tures, stress and strain, and the strength of ma-
      terials.

850. Föppl, August, and Ludwig Föppl. **Drang und
      Zwang, Eine höhere Festigkeitslehre für Inge-
      nieure.** Munich and Berlin: R. Oldenbourg,
      1920.

      Important work on stress and strain. August
      Föppl was the teacher of L. Prandtl and S.
      Timoshenko and father of Ludwig Foppl. Reprinted
      in 1969 by Johnson Reprint with an English intro-
      duction and brief biography of August Föppl by
      Gunhard Oravas.

851. Francis. James B. **On the Strength of Cast-Iron
      Pillars.** New York, 1865.

Presents a study of the strength of cast iron pillars but relies on European data.

852. Grashof, Franz. **Festigkeitslehre mit Rucksicht auf den Maschinenbau.** Berlin, 1866.

Textbook on the theory of elasticity by one of the founders of the Verein Deutscher Ingenieure and successor to Redtenbacher at Karlsruhe.

853. Hodgkinson, Eaton. "Experimental Researches on the Strength of Pillars of Cast Iron, and Other Materials." **Philosophical Transactions of the Royal Society of London** (1840): 385-457.

Describes research conducted with a machine capable of testing full-sized cast iron pillars. The machine had a capacity of over 22 tons.

854. _____. "On the Direct Tensile Strength of Cast Iron." **British Association for the Advancement of Science** Third Report (1833): 423-24.

Investigates the bending strength of cast iron. Shows that the neutral line shifts with increasing loads and allows the beam to bear greater loads than thought possible.

855. Kelvin, William Thomson, Baron. **Elasticity.** Edinburgh: A. & C. Black, 1878.

Based on an article written for the 1875 edition of the **Encyclopaedia Britannica.** Provides a survey of the theory of elasticity in the second half of the 19th century.

856. Lamé, Gabriel. **Leçons sur la théorie mathématique de l'élasticité des corps solides.** Paris: Bachelier, 1852.

Fundamental work on the theory of elasticity.

857. Maxwell, James Clerk. "On the Equilibrium of Elastic Solids." **Transactions of the Royal Society of Edinburgh** 20 (1853): 87-120.

Important paper on the theory of elasticity.

858.   Merriman, Mansfield. **A Text-Book on the Mechan-
        ics of Materials and of Beams, Columns, and
        Shafts.** New York: J. Wiley & Sons, 1885.

        Widely used textbook on the strength of mate-
        rials, based on applied mechanics, by a professor
        of civil engineering at Lehigh University. Went
        through numerous editions.

859.   Navier, Claude Louis Marie Henri. "Mémoire sur
        les lois de l'équilibre et du mouvement des
        corps solides élastiques." **Mémoires de
        l'Académie royale des sciences** 2nd series 8
        (1827): 375-93.

        Important work on the theory of elasticity by
        a leading French theoretician.

860.   Perronet, Jean Rodolphe. **Description des projets
        et de la construction des ponts de Neuilly,
        de Mantes, d'Orléans, & autres.** 2 volumes.
        Paris: L'Imprimerie royale, 1782-89.

        Examples of early attempts to apply scien-
        tific theory to bridge design and construction.

861.   Poisson, Siméon Denis. "Mémoire sur l'équilibre
        des corps élastiques." **Mémoires de
        l'Académie royale des sciences** 2nd series 8
        (1829): 357-570.

        Important paper on the theory of elasticity
        by a leading French mathematician and theoreti-
        cian.

862.   _____. "Mémoire sur les surfaces élastiques."
        **Mémoires del la Classe des sciences mathéma-
        tiques et physiques de l'Institut de France**
        13 (part 2) (1812): 167-225.

        Important paper on the modes of vibrations of
        elastic surfaces.

863.   Rankine, William John Macquorn. "Laws of the
        Elasticity of Solid Bodies." **The Cambridge
        and Dublin Mathematical Journal** 6 (1851):
        47-80.

        Important paper on the theory of elasticity.
        One of the first to introduce the modern concept
        of stress and strain into English.

864.   Reech, Ferdinand (Frédéric). **Cours de mécanique d'après la nature généralemant flexible et élastique des corps, comprenant la statique et la dynamique avec la théorie des vitesses virtuelles.** Paris: Carilian-Goeury, 1842.

        Theory of elasticity based on Newtonian principles.

865.   Rennie, George. "Account of Experiments Made on the Strength of Materials." **Philosophical Magazine** 53 (1819): 161-75.

        Describes experiments made with a machine designed to test samples in either tension or compression. Discovers that there is no simple relationship between the strength of stones and their specific gravity or hardness. Results were useful to bridge engineers.

866.   Robison, John. "Strength of Materials." **Encyclopaedia Britannica.** Third Edition, 1798.

        Describes 18th-century work on the strength of materials, including that of P. Musschenbroek.

867.   Rogers, William Barton. **An Elementary Treatise on the Strength of Materials.** Charlottesville, Va., 1838.

        One of the first American books to present a mathematical theory of the properties of materials. Produced as part of an effort to create an engineering school at the University of Virginia. Later Barton was associated with the founding of M.I.T.

868.   Saint-Venant, Adhémar Jean Claude Barré de. **The Elastic Researches of Barré de Saint-Venant.** Edited by Karl Pearson. Cambridge: Cambridge University Press, 1889.

        Includes fundamental research on elasticity, torsion, and flexure by a leading figure in the field.

869.   _____. **Mémoire sur la flexion des prismes, sur les glissements transversaux et longitudinaux qui l'accompagnent lorsqu'elle ne s'opère pas uniformément ou en arc de cercel, et sur la**

forme courbe affectée alors par leurs sections transversales priniticement planes.
Paris: Mellet-Bachelier, 1856.

Fundamental work on the theory of flexion.

870.   _____. Mémoires sur la résistance des solides
        suivis d'une note sur la flexion des pièces à
        double courbure. Paris: Bachelier, 1844.

        Important works on the theory of elasticity.
        Extracted from five articles published in the
        Comptes rendus des séances de l'Académie des sci-
        ences.

871.   _____. "Mémoire sur la torsion des prismes,
        avec des considerations sur leur flexion,
        ainsi que l'équilibre intérieur des solides
        élastiques en général, et des formules pra-
        tiques pour le calcul de leur résistance à
        divers efforts s'exerçant simultanément."
        Mémoires de l'Académie des sciences series 2
        14 (1856): 253-560.

        Fundamental paper on the theory of elastic-
        ity.

872.   Thurston, Robert Henry. "Flexure of Beams."
        Transactions of the American Society of Civil
        Engineers 4 (1875): 284-99.

        Studies of stress and strain in beams by a
        leading American engineering scientist.

873.   _____. The Materials of Engineering. 3 vol-
        umes. New York: J. Wiley & sons, 1883-84.

        Presents a study of the strength of materials
        based on experimentation. Represents a new, ma-
        ture level of American work on the strength of
        materials. Also published in an abridged form as
        A Text-Book of the Materials of Construction, for
        Use in Technical and Engineering Schools. New
        York: J. Wiley & Sons, 1890.

874.   _____. "On the Strength, Elasticity, Ductility
        and Resilience of Materials of Machine Con-
        struction, and on Various Hitherto Unobserved
        Phenomena Noticed During Experimental Re-
        searches with a New Testing Machine Fitted
        with an Autographic Registry." Transactions

of the American Society of Civil Engineers 2 (1874): 349-78.

Presents experimental studies of the strength of materials.

875. Tredgold, Thomas. "On the Resistance of Solids; with Tables of the Specific Cohesion and Cohesive Force of Bodies." **Philosophical Magazine** 50 (1817): 413-28.

Surveys the current state of knowledge of structural materials. Applies geometrical arguments, such as the neutral line between tension and compression, to experimental tests.

876. _____. **Practical Essay on the Strength of Cast Iron and Other Metals.** London: J. Taylor, 1824.

Early work on the strength of iron based on experimental testing. Translated into French, German and Italian. Went through many editions.

877. U.S. Ordinance Department. **Report of Experiments on the Strength and Other Properties of Metals for Cannon, with a Description of the Machines for Testing Metals.** Philadelphia, 1856.

Presents some of the first experiments conducted in America on the strength of materials. Work done by Maj. William Wade and Capt. Thomas Jackson Rodman. First American tests to attract European interest. Describes one of the earliest testing machines to be built in America.

878. _____. **Reports of Experiments on the Properties of Metals for Cannon, and the Qualities of Cannon Powder.** Boston, 1861.

Presents the results of experimental tests done by Capt. T.J. Rodman. Gives empirical methods for solving problems in the strength of materials using pressure gauges and models.

879. Winkler, Emil. **Die lehre von der elasticität und festigkeit mit besonderer rucksicht auf ihre anwendung in der technik, für polytechnische schulen, bauakademien, ingenieure,**

maschinenbauer, architecten. Prague: H.
Dominicus, 1867.

Work on the theory of elasticity by a pioneering engineering scientist.

880. Wood, De Volson. **A Treatise on the Resistance of
       Materials, and an Appendix on the Preserva-
       tion of Timber.** New York: J. Wiley & Son,
       1871.

Popular American textbook on the strength of
materials. Based on lectures given to the senior
civil engineering class at the University of
Michigan. Went through several editions.

881. Zimmermann, Hermann. **Die Knickfestigkeit eines
       Stabes mit elastischer Querstutzung.** Berlin:
       W. Ernst, 1906.

Application of graphical stress analysis by a
leading structural engineer.

See also: 387, 388, 393, 401, 411, 601, 690, 691, 709,
716, 732, 744, 746, 755, 759, 775, 790, 819, 904, 907,
930, 1086.

Secondary Sources

882. Baumann, R. "Das Materialprüfungswesen und die
       Erweiterung der Erkenntrisse auf Gebiet der
       Elastizität und Festigkeit in Deutschland
       während der letzten vier Jahrzehnte."
       **Beiträge zür Geschichte der Technik und In-
       dustrie: Jahrbuch des Vereines Deutscher In-
       genieure** 4 (1912): 147-95.

Discusses the history of materials testing in
Germany.

883. Boulingard, Georges. "La mécanique théorique des
       corps flexibles (1638-1788) et les premières
       tentatives de 'spéculations fonctionelles' au
       XVIIIe siècle." **Revue d' Histoire des Sci-
       ences** 17 (1964): 13-24.

Discusses 18th-century theories of elasticity.

884. Cohen, Morris, ed. **Materials Science and Engineering: Its Evolution, Practice and Prospects.** Lausanne: Elsevier Sequoia, 1979.

Work is divided into two parts: (1) "Materials in History and Society" by Melvin Kranzberg and Cyril Stanley Smith, and (2) "Materials Science and Engineering as a Multidiscipline" by Richard Chassen and Alan G. Chynoweth. Part 1 emphasizes the role of metals in materials science but also deals with ceramics and general solid state structures. Argues that the use of materials requires an interaction between physics, chemistry and engineering. Part 2 focuses on ten case histories of modern materials science.

885. Gibbons, Chester H. "History of Testing Machines for Materials." **Transactions of the Newcomen Society** 15 (1934-35): 169-84.

Provides a brief survey of the development of testing machines.

886. _____. **Materials Testing Machines.** Pittsburgh: Instruments Publishing Co., 1935.

Provides some useful information on the history of materials testing and the role of testing machines in the development of a theory of the strength of materials.

887. Girill, T.R. "The First Law of Elasticity." **American Journal of Physics** 40 (1972): 16-20.

Investigates the relation between the gas law of Boyle and Mariotte, which was based on pneumatic springs, and Hooke's stress-strain relationship. Shows how Hooke's work became disassociated from the work on gases.

888. Mark, Robert, and James K. Chiu, and John F. Abel. "Stress Analysis of Historic Structures: Maillart's Warehouse at Chiasso." **Technology and Culture** 15 (1974): 49-63.

Uses modern photoelastic stress analysis to study the 1924 warehouse by architect and bridge builder Robert Maillart. Discovers connections to Maillart's light bridge designs. Concludes

that the frame of the shed is more rational than
architectural historians have understood. Provides a new method for understanding historic
structures.

889. Markovitz, Herschel. "The Emergence of Rheol-
          ogy." **Physics Today** 21 (April 1968): 23-30.

      Traces the history of rheology to Newton and
Hooke.

890. Oravas, G.A., and L. McLean. "Historical Devel-
          opment of Energetical Principles in Elastome-
          chanics. From Heraclitos to Maxwell." **Ap-
          plied Mechanics Review** 19 (1966): 647-58.

      Traces the history of elastomechanics. In-
cludes a discussion of fluid mechanics.

891. Petik, Ferenc. "The Development of Material
          Testing Machines." **Technikatorteneti Szemle**
          11 (1979): 217-32.

      Discusses the history of experimental materi-
als testing.

892. Smith, A.I. "William Fairbairn and the Mechani-
          cal Properties of Materials." **The Engineer**
          217 (26 June 1964): 1133-36.

      Discusses Fairbairn's studies of fatigue.

893. Steiner, Frances. "Building with Iron: A
          Napoleonic Controversy." **Technology and Cul-
          ture** 22 (1981): 700-724.

      Argues that Napoleon's interest in iron as a
building material gave power to engineers over
architects. Believes that his interest in iron
can be traced to his exposure to a scientific approach to engineering at the École Militaire.

894. Sutherland, R.J.M. "The Introduction of Struc-
          tural Wrought Iron." **Transactions of the
          Newcomen Society** 36 (1963-64): 67-82.

      Shows how the needs of the railways for
stronger bridges led to the introduction of
wrought iron as a structural material. Discusses
the work of E. Hodgkinson and W. Fairbairn on the
testing of wrought iron.

895. Szabo, Istvan. "Die Grundlege der linearen Elas-
      tizitätstheorie für homogene und isotrope
      Körpe." **Technikgeschichte** 40 (1973): 301-
      36.

      Surveys the theory of elasticity from Galileo
      to Cauchy.

896. Todhunter, Isaac. **A History of the Theory of
      Elasticity and of the Strength of Materials,
      from Galileo to the Present.** 2 volumes.
      Edited by Karl Pearson. Cambridge: Cambridge
      University Press, 1886-1893.

      Provides an extensive study of the history of
      the theory of elasticity. Contains detailed
      analysis of the various theories of elasticity.
      Organized around individual theorists. Useful
      source book. Reprinted by Dover Publications.

897. Timoshenko, Stephen P. **History of Strength of
      Materials.** New York: McGraw-Hill, 1953.

      Provides a comprehensive descriptive study of
      the history of the strength of materials. In-
      cludes a useful discussion on the experimental
      testing of materials and testing machines.

898. Truesdell, Clifford A. "Outline of the History
      of Flexible or Elastic Bodies to 1788."
      Journal of the Acoustical Society 32 (1960):
      1647-56.

      Surveys the early history of the theory of
      elasticity.

899. _____. "The Rational Mechanics of Materials--
      Past, Present, Future." **Applied Mechanics
      Review** 12 (1959): 75-80.

      Includes a discussion of the history of elas-
      ticity.

900. _____. "The Rational Mechanics of Flexible and
      Elastic Bodies, 1638-1788." **Leonhardi Euleri
      Opera Omnia.** 2nd series 11(2) (1960): 7-
      435.

Detailed analysis of the development of the theory of elasticity during the 17th and 18th centuries.

901.   _____. "Zür Geschichte des Begriffes 'innerer Druck.'" **Physikalische Blätter** 12 (1956): 315-26.

Discusses the development of the concept of stress. Shows how the concept was connected to work in fluid dynamics.

902.   Unwin, William. "The Experimental Study of the Mechanical Properties of Materials." **Institution of Mechanical Engineers** (October, 1918): 405-440.

Describes the history of materials testing. Includes a discussion of Telford's tests done using the chain testing machine at Woolwich Dockyard.

See also: 76, 202, 237, 293, 301, 311, 328, 331, 332, 376, 565, 678, 802, 818, 831, 833, 956, 958, 962.

THE THEORY OF STRUCTURES AND STATICS

Primary Sources

903.   Airey, George Bibbell. "On the Use of the Suspension Bridge with Stiffened Roadway, for Railway and other Bridges of Great Span." **Proceedings of the Institution of Civil Engineers** 26 (1867): 258-64.

Provides an analysis of suspension bridges based on experimentation.

904.   Barlow, Peter. "On the Mechanical Effect of Combining Girders and Suspension Chains, and a Comparison of the Weight of Metal in Ordinary and Suspension Girders to Provide Equal Deflections with a Given Load." **British Association for the Advancement of Science Report** (1857), pp. 238-48.

Describes experimental tests that were used as part of the analysis of the design of bridges.

905.   Belidor, Bernard Forest de.   **La science des ingé-
       nieurs dans la conduite des travaux de forti-
       fication et d'architecture civile.**   New edi-
       tion with notes by C. Navier.   Paris: F.
       Didot, 1813.

       Fundamental 18th-century work on the applica-
       tion of science to fortifications and buildings.
       Nineteenth-century edition with notes by Navier.
       One of the first works to use the term engineer-
       ing science.   Originally published in Paris by C.
       Jombert in 1729.

906.   Borgnis, Guiseppe Antonio (Joseph A.).   **Elementi
       di statica architettonica.**   Milan: G. Truffi,
       1842.

       Textbook on the principles of statics in
       structures.

907.   _____.   **Traité élémentaire de construction ap-
       pliquée à l'architecture civile.**   Paris:
       Bachelier, 1823.

       Treatise on the theory of structures.   In-
       cludes a section on materials.

908.   Bossut, Charles, and Guillaume Viallet.
       **Récherches sur la construction la plus avan-
       tageuse des digues.**   Paris: C.A. Jombert,
       1764.

       Early research on the theory of dams and
       dikes.   Based on a prize competition by the
       Académie des Sciences in 1762.

909.   Fairbairn, Sir William.   **An Account of the Con-
       struction of the Britannia and Conway Tubular
       Bridges, with a Complete History of Their
       Progress, from the Conception of the Original
       Idea, to the Conclusion of the Elaborate Ex-
       periments which Determined the Exact Form and
       Mode of Construction Ultimately Adopted.**
       London: J. Weale, 1849.

       Describes the role of experiments in the con-
       struction of the Britannia and Conway bridges.

910.   _____.   **On the Application of Cast and Wrought
       Iron to Building Purposes.**   London: J. Weale,
       1854.

Went through several editions and includes a
discussion of iron bridge design.

911. Gauthey, Émil and Marie. **Mémoire sur
      l'application des principes de la mécanique à
      la construction des voutes et domes, dans
      lequel on examine le probleme proposé par M.
      Patte, relativement à la construction de la
      couple de l'Eglise Sainte-Geneviéve de Paris.**
      Paris: C.A. Jombert, 1771.

      Early work on the application of the princi-
      ples of mechanics to the construction of vaults
      and domes.

912. _____. **Traité de la construction des ponts.** 3
      volumes. Paris: F. Didot, 1809-16.

      Important French work on the theory of
      bridges. Edited by C.L.M.H. Navier.

913. Greene, Charles E. **Graphics for Engineers, Ar-
      chitects, and Builders.** New York: John Wiley
      & Sons, 1879.

      Application of graphical methods to struc-
      tures. Written by a professor of civil engineer-
      ing at the University of Michigan. Divided into
      three parts: I Roof-Trusses; II Bridge Trusses;
      and III Arches, in Wood, Iron and Stone.

914. Haupt, Herman. **General Theory of Bridge Con-
      struction.** New York: D. Appleton, 1851.

      Presents an American attempt to develop a
      mathematical theory of bridges based on European
      theory. Relies on the scientific theories of
      Thomas Young. Assumes, wrongly, that stresses
      could be reduced to forces between particles.
      Uses graphical methods. Went through several
      editions.

915. _____. **Military Bridges: With Suggestions of
      New Expedients and Constructions for Crossing
      Streams and Chasms; Including, also, Designs
      for Trestle and Truss Bridges for Military
      Railroads, Adapted Especially to the Wants of
      the Service in the United States.** New York:
      D. Van Nostrand, 1864

Based on his **General Theory of Bridge Construction** (item 914) but also containing experimental work on blanket boats.

916.   Hutton, Charles. **The Principles of Bridges: Containing the Mathematical Demonstrations of the Properties of the Arches, the Thickness of the Piers, the Force of the Water Against Them, Etc.** Newcastle: Saint, 1772.

Important early work on the mathematical theory of bridges.

917.   _____. **Tracts on Mathematical and Philosophical Subjects; Comprising, Among Numerous Important Articles, the Theory of Bridges, with Several Plans of Recent Improvements.** 3 volumes. London: F.C. and J. Rivington, 1812.

Contains early work on the theory of bridge design. Includes the results of experiments.

918.   Jenkin, Henry Charles Fleeming. **Bridges: An Elementary Treatise on Their Construction And History.** Edinburgh: A. & C. Black, 1876.

Work on the theory of bridges by the professor of engineering at the University of Edinburgh. Based on an article in the **Encyclopaedia Britannica.**

919.   _____. "On the Practical Application of Reciprocal Figures to the Calculation of Strains on Framework." **Transactions of the Royal Society of Edinburgh** 25 (1869): 444-447.

Paper on the theory of structures.

920.   Mahan, Dennis H. **An Elementary Course of Civil Engineering.** New York: Wiley and Putnam, 1837.

First American textbook on structures to be based on French engineering theory. Used at West Point; also had a large impact at other schools. Went through several editions.

921.   _____. **An Elementary Course of Military Engineering.** New York: J. Wiley, 1865.

Applies French theory to military fortifications. Used at West Point.

922. _____. **A Treatise on Civil Engineering.** Revised and edited by De Volson Wood. New
York: John Wiley & Sons, 1873.

Work on civil engineering and the theory of
structures based on French engineering theory.
Used at West Point and other schools.

923. Maxwell, James Clerk. "On Reciprocal Figures and
Diagrams of Forces." **Philosophical Magazine**
27 (1864): 250-61.

Important study on the use of reciprocal figures to study the forces acting in structures.
Precipitated by W.J.M. Rankine's idea of parallel
projection.

924. _____. "On Reciprocal Figures, Frames, and Diagrams of Forces." **Transactions of the Royal
Society of Edinburgh** 26 (1872): 1-40.

Further study on reciprocal figures.

925. _____. "On the Calculation of the Equilibrium
and Stiffness of Frames." **Philosophical Magazine** 27 (1864): 294-99.

Mathematical analysis of the theory of
frames.

926. Merriman, Mansfield. **A Text-Book on Retaining
Walls and Masonry Dams.** New York: John Wiley
& Sons, 1892.

Textbook by a professor of civil engineering
at Lehigh University.

927. _____, and Henry S. Jacoby. **A Text-Book on
Roofs and Bridges.** New York: John Wiley &
Sons, 1888-98.

Contains four parts published separately.
Part I: Stresses in Simple Trusses; Part II:
Graphic Statics; Part III: Bridge Design; Part
IV: Higher Structures. Merriman was professor
of civil engineering at Lehigh University and
Jacoby was professor of bridge engineering at
Cornell. Important work.

928.   Monge, Gaspard. **Traité élémentaire de statique.**
        Paris, 1788.

        Textbook by one of the founders of the École
        Polytechnique on the use of geometry to analyze
        static forces. Later revised by Jean Nicolas
        Pierre Hâchette in 1810.

929.   Navier, Claude Louis Marie Henri. **Rapport à Monsieur Becquey ... director général des ponts
        et chaussées et des mines, et Mémoire sur les
        ponts suspendus.** Paris: L'imprimerie royale,
        1823.

        Mathematical study of suspension bridges by a
        leading French theoretician.

930.   Newlon, Howard, Jr., ed. **A Selection of Historic
        American Papers on Concrete, 1876-1926.**
        Detroit: American Concrete Institute, 1976.

        Reprints seven landmark papers on the devel-
        opment of reinforced concrete construction. In-
        cludes articles on the experimental testing of
        concrete beams and slab floors.

931.   Poncelet, Jean Victor. **Mémoire sur la stabilité
        des révêtements et de leurs foundations.**
        Paris: Bachelier, 1840.

        Mathematical study of the stability of revet-
        ments and their foundations, conducted by a lead-
        ing French engineering scientist and professor of
        mechanics applied to machines at the École
        d'Application de l'Artillerie et du Génie at
        Metz.

932.   Rankine, William John Macquorn. **A Manual of
        Civil Engineering.** London: Charles Griffin &
        Co., 1862.

        Treatise on civil engineering based on the
        theory of applied mechanics. Treats the theory
        of structures and earthworks. Widely used as a
        textbook in university-level engineering courses.
        Went through several editions.

933. Rennie, George. **An Historical, Practical and Theoretical Account of the Breakwater in Plymouth Sound.** London: H.G. Bohn, and J. Weale, 1848.

Provides a theoretical analysis of the Plymouth sound breakwater.

934. _____. **The Theory, Formation, and Construction of British and Foreign Harbours.** London: J. Weale, 1854.

Puts forward a theoretical analysis of harbors.

935. Robison, John. "Carpentry." **Encyclopaedia Britannica.** Third Edition Supplement, 1803.

Useful overview of the theory of wooden structures at the beginning of the 19th century.

936. Smeaton, John. **A Narrative of the Building and a Description of the Construction of the Edystone Lighthouse with Stone.** London: G. Nicol, 1791.

Describes the design of the Edystone lighthouse and the role of biological models in its design.

937. Tredgold, Thomas. **Elementary Principles of Carpentry.** London: J. Taylor, 1820.

Early pioneering work on the strength of wood and applied mechanics. Went through numerous editions.

938. Weale, John, ed. **The Theory, Practice, and Architecture of Bridges of Stone, Iron, Timber, and Wire; with Examples on the Principle of Suspension.** London: Architectural Library, 1843.

Contains essays on the theory of bridge design. Includes a paper on the theory of the arch by Henry Moseley.

939. Whipple, Squire. **An Elementary and Practical Treatise on Bridge Building.** Utica, N.Y.: H.A. Curtis, 1847.

Early American work on bridge design using only elementary mathematics. Presents graphical and mathematical methods for the analysis of truss bridges which could not be analyzed by other methods. Whipple's work was independent of European studies. Republished in 1873.

940.   Winkler, Emil. **Vortrage über Bruckenbau.** 4 vols. Vienna: C. Gerold, 1872-81.

Treatise on bridge building by a leading engineering scientist.

941.   Wood, De Volson. **Treatise on the Theory of the Construction of Bridges and Roofs.** New York: J. Wiley & Son, 1873.

Important work on American bridge-building techniques. Used as a textbook. Went through several editions.

942.   Young, Thomas. "Bridges." **Encyclopaedia Britannica.** 7th edition, 1842.

Surveys the knowledge of the theory of bridges during the middle of the 19th century.

943.   _____. **Treatise on Masonry, Joinery, and Carpentry.** Edinburgh: A. & C. Black, 1839.

Survey of the theory of structures based on articles in the 7th edition of the **Encyclopaedia Britannica.**

See also: 401, 406, 698, 724, 739, 743, 746, 775, 792, 793, 819, 849, 860.

Secondary Sources

944.   Addis, W. "A New Approach to the History of Structural Engineering." **History of Technology** 8 (1983): 1-13.

Focuses on the role of engineering design in the history of structural engineering. Argues that the concept of design would overcome the old distinction between theory and practice.

945.   Billington, David P. "Bridges and the New Art of
        Structural Engineering." **American Scientist**
        72 (January-February 1984): 22-31.

        Discusses the role of art and science in the
        design of bridges.

946.   _____. "Engineering Education and the Origins
        of Modern Structures." **Civil Engineering** 39
        (January 1969): 52-57.

        Argues in favor of examining the 19th-century
        origins of modern structures.

947.   _____. "An Example of Structural Art: The
        Salginatobel Bridge of Robert Maillart."
        **Journal of the Society of Architectural
        Historians** 33 (March 1974): 61-72.

        Includes a technical analysis of Maillart's
        bridge design.

948.   _____. **Robert Maillart's Bridges: The Art of
        Engineering**. Princeton: Princeton University
        Press, 1979.

        Describes the work of the Swiss structural
        engineer Robert Maillart. Argues that his stud-
        ies of reinforced concrete led to new ideas of
        bridge design based on the union of scientific
        and aesthetic principles. Provides a systematic
        analysis of Maillart's bridges. Argues for the
        role of intuition and imagination, along with
        scientific analysis, in order to obtain elegance
        in design. Important work.

949.   _____. **The Tower and the Bridge: The New Art of
        Structural Engineering**. New York: Basic
        Books, 1983.

        Argues for the idea of "structural art" as
        distinct from science, architecture and sculp-
        ture. Believes that successful structural art
        such as the Eiffel Tower and the Brooklyn Bridge
        have three qualities: efficiency, economy, and
        elegance. Concludes that the first quality in-
        volves the relationship between engineering and
        science, but the others involve ideas of art and
        design.

950. Charlton, T.M. "Contributions to the Science of
       Bridge-Building in the 19th Century by Henry
       Moseley, Hon. LL.D., F.R.S. and William Pole,
       D. Mus., F.R.S." **Notes and Records of the
       Royal Society of London** 30 (1976): 169-79.

       Discusses Moseley's and Pole's contribution
       to structural theory.

951. _____. **A History of Theory of Structures in the
       Nineteenth Century**. New York: Cambridge Uni-
       versity Press, 1982.

       Internalist approach with little biographical
       material. Provides a straightforward history of
       technical theory but with few connections between
       different structures. Places special emphasis on
       the role of science.

952. _____. "Maxwell, Jenkin and Cotterill and the
       Theory of Statically-Indeterminate Struc-
       tures." **Notes and Records of the Royal Soci-
       ety of London** 26 (1971): 233-46.

       Discusses the development of a theory of in-
       determinate structures.

953. Collins, A.R. **Structural Engineering: Two Cen-
       turies of British Achievement**. Chislehurst,
       Kent: Tarot Print, 1983.

       Discusses the British contributions to struc-
       tures and structural theory.

954. Condit, Carl W. **American Building Art: The Nine-
       teenth Century**. New York: Oxford University
       Press, 1960.

       Discusses the transition of building art from
       a period of empiricism to one based on scientific
       and mathematical theories. Covers wood framing,
       iron framing, wooden bridge trusses, iron bridge
       trusses, suspension bridges, iron arch bridges,
       railway trainsheds, and concrete construction.
       Important work.

955. _____. **American Building Art: The Twentieth
       Century**. New York: Oxford University Press,
       1961.

Discusses the role of new structural theory
and the development of 20th-century architecture.
Concludes that we have failed to coordinate architecture and engineering.

956.   _____. **American Building: Materials and Tech-
        niques from the First Colonial Settlements to
        the Present.** Chicago: University of Chicago
        Press, 1968.

Discusses the change in building techniques
from craft to modern engineering. Discusses the
calculation of stress, the scientific study of
materials, and new structural forms. Summarizes
in one volume many of the ideas in his books on
American Building Art (items 954, 955).

957.   _____. **The Chicago School of Architecture.**
        Chicago: University of Chicago Press, 1964.

Discusses the work of Louis Sullivan, J.W.
Root, D.H. Burnham and D. Adler. Includes a discussion of the role of structural theories in
their work.

958.   _____. "The First Reinforced-Concrete
        Skyscraper: The Ingalls Building in Cincin-
        nati and Its Place in Structural History."
        Technology and Culture 9 (1968): 1-33.

Includes a discussion of the evolution of
concrete construction as a scientific technology.
Discusses the role of industrial testing. Shows
how the Ingalls Building incorporated modern scientific reinforcing systems.

959.   _____. "The Structural System of Adler and
        Sullivan's Garrick Theater Building."
        Technology and Culture 5 (1964): 523-40.

Demonstrates that the structural system of
the Garrick Theater Building combines a scientific approach to structures along with an empirical pragmatic approach.

960.   _____. "Sullivan's Skyscrapers as the Expres-
        sion of Nineteenth Century Technology."
        Technology and Culture 1 (1959): 78-93.

Discusses how the transformation of building
from an empirical technique to an exact science

influenced the work and thought of Louis
Sullivan. Includes a discussion of the role of
science in the Eads Bridge.

961. Cowan, Henry J. **A Historical Outline of Archi-
      tectural Science.** Amsterdam: Elsevier
      Publishing, 1966.

      Provides a technical background for an under-
      standing of structural theory. Treats the sub-
      ject more topically than historically. Aimed
      more at architects than historians.

962. _____. **Science and Building: Structural and En-
      vironmental Design in the Nineteenth and
      Twentieth Century.** New York: John Wiley &
      Sons, 1978.

      Discusses the role of structural theory and
      the science of materials in the design of build-
      ings. Discusses the use of analogues, models,
      and computers in the process of design as well as
      the development of mathematical theories of
      structures. Provides a great number of facts but
      lacks a sufficient historiographic framework.

963. Danko, George M. "The Evolution of the Simple
      Truss Bridge, 1790-1850: From Empiricism to
      Scientific Construction." Ph.D. disserta-
      tion. University of Pennsylvania, 1979.

      Describes the scientific, technical, social
      and economic context of the evolution of the
      truss bridge in America. Analyzes the role of
      scientific and technical knowledge and the rise
      of the civil engineering profession in the design
      of bridges. Discusses the influence of European
      technical literature on American engineers.
      Shows how the role of experiments led to iron re-
      placing wood in bridge design. Concludes that
      new college educated designers replaced empirical
      builders as the cost of iron bridges needed for
      the railroad began to increase.

964. Dorn, Harold I. "The Art of Building and the
      Science of Mechanics: A Study of the Union of
      Theory and Practice in the Early History of
      Structural Analysis in England." Ph.D. dis-
      sertation. Princeton University, 1970.

Discusses the role of science in the development of structural theory. Includes an analysis
of the role of W.J.M. Rankine's engineering
science.

965.   _____. "A Note on 'Structural Antecedents of
        the I-Beam.'" **Technology and Culture** 9
        (1968): 415-18.

Provides a response to an article by Robert
Jewett (item 971). Argues that it is not quite
accurate to say that the I-beam was developed out
of empirical methods and practice. Concludes
that Tredgold's work was theoretical. Followed
by a response from Jewett (pp. 419-26) and a
rejoinder by Dorn (pp. 427-29).

966.   Hamilton, Stanley B. "The Historical Development
        of Structural Theory." **Proceedings of the
        Institution of Civil Engineers**   (1952): 374-
        419.

Surveys the historical development of structural theory.

967.   Hertwig, A. "Die Entwicklung der Statik der
        Baukonstruktionen im 19. Jahrhundert." **Tech-
        nikgeschichte** 30 (1941): 82-98.

Surveys the history of the application of
statics to construction during the 19th century.

968.   Hopkins, H.J. **A Span of Bridges: An Illustrated
        History**. New York: Frederick A. Praeger,
        1970.

Presents an early written account of the history of bridges. Includes sections on the
strength of materials, structural analysis, and
mechanics. Organized into three parts: compression bridges, such as stone arches; tension
bridges, such as truss and suspension bridges;
and tension and compression bridges, such as concrete bridges.

969.   James, J.G. "The Evolution of Iron Bridge
        Trusses to 1850." **Transactions of the New-
        comen Society** 52 (1980-81): 67-101.

Discusses the truss as a structural element
in iron bridges.

970.   _____. "Iron Arched Bridge Designs in Pre-Revo-
        lutionary France." **History of Technology** 4
        (1979): 63-99.

        Discusses the interest in iron bridges in
        18th century France. Shows the influence on
        English designs, models and published proposals.
        Describes new innovations in structural design.

971.   Jewett, Robert A. "Structural Antecedents of the
        I-Beam, 1800-1850." **Technology and Culture**
        8 (1967): 346-62.

        Traces the history of the development of the
        I-shape as an element of structures. Argues that
        structural theory contributed little to the emer-
        gence of the I-shape. Argues that the I-beam
        arose out of rolling-mill requirements and limi-
        tations, and out of engineering practice and ex-
        perimentation. Shows that the I-shape arose from
        practical problems in building floor beams, rail-
        road rails, iron ship framing and bridge girders.
        For a response by Harold Dorn, see item 965.

972.   Kemp, E.L. "Links in a Chain: The Development
        of Suspension Bridges, 1801-70." **Structural
        Engineer** 57A (August 1979): 255-63.

        Argues that the modern suspension bridge was
        made possible by the introduction of new building
        materials and techniques.

973.   Kouwenhoven, John A. "The Designing of the Eads
        Bridge." **Technology and Culture** 23 (1982):
        535-68.

        Discusses the design and construction of the
        Eads bridge at St. Louis.

974.   Kranakis, Eda Fowlks. "Technological Styles in
        America and France in the Early 19th Century:
        The Case of the Suspension Bridge." Ph.D.
        dissertation. University of Minnesota, 1982.

        Discusses the difference between American and
        French approaches to technology through a compar-
        ison of suspension bridges. Distinguishes between
        a mathematical deductive approach in France and
        an experimental inductive approach in America.

975. McCullough, David. **The Great Bridge.** New York:
         Simon & Schuster, 1972.

         Provides a study of the design and construc-
      tion of the Brooklyn Bridge.

976. Mainstone, Rowland J. **Developments in Structural
         Form.** Cambridge, Mass.: M.I.T. Press, 1975.

         Presents a broad survey of the history of
      building techniques. Organized analytically and
      historically into four parts: (1) structural the-
      ory, (2) structural elements, (3) completed
      structures, (4) history of structural theory.
      Analyzes the relationship of structural form to
      material and physical properties.

977. Miller, Howard S., and Quinta Scott. **The Eads
         Bridge.** Columbia, Mo.: University of
         Missouri Press, 1979.

         Includes a photographic essay by Scott and an
      historical analysis by Miller of the famous
      bridge at St. Louis. Argues that Eads' engineer-
      ing was based on the belief that God created a
      rational world which could be mastered by tech-
      nology. Discusses the influence of Telford's de-
      signs on Eads. Discusses the structural problem
      of designing the first steel bridge.

978. Pallett, J.E. "The Contribution of William
         Froude to the Development of the Oblique
         Bridge with Mechanically Correct Spiral Taped
         Courses." **Transactions of the Newcomen Soci-
         ety** 45 (1972-73): 205-215.

         Discusses the contributions to structural
      theory by Froude, who was more well known for his
      work in naval architecture.

979. Perez-Gomez, Alberto. **Architecture and the Cri-
         sis of Modern Science.** Cambridge, Mass.:
         M.I.T. Press, 1983.

         Argues that since the time of Descartes, sci-
      ence has had a negative impact on architecture.

980. Petterson, Charles T. "Inventing the I-Beam:
         Richard Turner, Cooper & Hewitt, and Others."
         **Bulletin of the Association for Preservation
         Technology** 12, no. 4 (1980): 3-28.

Discusses the development of an important element in the history of structures.

981.   Plowden, David. **Bridges: The Spans of North America.** New York: Viking Press, 1974.

Presents a comprehensive study of over 650 bridges. Organized according to the materials used, such as stone, wood, iron, steel and concrete; consequently, presents different types of bridges, such as arch, truss, or suspension, under a single chapter on iron or steel. Relatively little on the development of structural theory. Emphasizes the description of the bridges and the biographies of bridge builders.

982.   Richardson, George. "19th Century Bridge Design in Canada: A Technology in Transition." **History of Science and Technology in Canada Bulletin** (September 1981): 177–86.

Discusses structural theory in Canada.

983.   Ruddock, Ted. **Arch Bridges and Their Builders, 1735–1835.** Cambridge: Cambridge University Press, 1979.

Describes the development of the application of scientific theory to bridge design from the beginning of the Westminster Bridge to the bridges of Thomas Telford and John Rennie. Argues that the development of scientific theory was disjointed. Distinguishes between the roles of the architect and the engineer. Documents the influence of French theory on British bridge design.

984.   Salvadori, Mario, and Robert Heller. **Structure in Architecture.** Englewood Cliffs, N.J.: Prentice-Hall, 1967.

Provides a non-mathematical discussion of the scientific theories of structures. Discusses loads, stresses and structural materials. Useful for understanding the technical basis of contemporary building technology.

985. Smith, Denis. "Structural Model Testing and the
          Design of British Railway Bridges in the
          Nineteenth Century." **Transactions of the
          Newcomen Society** 48 (1976-77): 73-90.

          Argues that the problem of load and vibration
      resulting from the invention of the steam locomo-
      tive made traditional design inadequate. Con-
      cludes that the problem led to the use of models
      for testing new designs.

986. _____. "The Use of Models in Nineteenth Century
          British Suspension Design." **History of Tech-
          nology** 2 (1977): 169-214.

          Shows that the use of models as an aid to de-
      sign and evaluation was a significant contribu-
      tion of British engineers during the 19th cen-
      tury. Describes the test apparatus of Thomas
      Telford and the experiments of Peter Barlow and
      G.B. Airey. Compares the use of models to the
      French use of mathematical analysis and the Amer-
      ican use of trial and error.

987. Smith, Norman. **A History of Dams.** Secaucus,
          N.J.: Citadel Press, 1972.

          Surveys the entire history of the construc-
      tion of dams from ancient times to the present.
      Provides a quantitative analysis of the dams.
      Analyzes the change in form of dams brought about
      by the application of theoretical and experimen-
      tal science during the 19th century. Discusses
      the relationship between form and function and
      the relationship between form and the materials
      available for dam construction.

988. Torroja, Edvardo. **Philosophy of Structures.**
          English version by J.J. and Milos Polivka.
          Berkeley: University of California Press,
          1958.

          Presents the ideas of a great Spanish
      builder. Discusses the role of modern structural
      analysis in the design of buildings.

989. Tyson, S. "Notes on the History, Development,
          and Use of Tubes in the Construction of
          Bridges." **Industrial Archaeology Review** 2
          (1978): 143-53.

Reviews the use of tubes as a structural element in bridge design.

990. Vincenti, Walter G., and Nathan Rosenberg. **The Britannia Bridge: The Generation and Diffusion of Technological Knowledge.** Cambridge, Mass.: M.I.T. Press, 1978.

Describes the contributions of Robert Stephenson, William Fairbairn and Eaton Hodgkinson to the design of a tubular bridge for the railroad. Discusses the applications of scientific knowledge, including models and the theory of beams, to the design of the bridge. Analyzes how technological knowledge is gained and diffused. Important work.

991. Wachsmann, Konrad. **The Turning Point of Building: Structure and Design.** Translated by Thomas E. Burton. New York: Reinhold, 1961.

Describes the change in building from handicraft skill to scientific design. Discusses historical examples such as the Crystal Palace, the Eiffel Tower, the Brooklyn Bridge, and the Firth of Forth Railway Bridge. Argues for building to be totally predicated on scientific design.

992. Witthoft, Hans. **Building Bridges: History, Technology, Construction.** Translated by Edward Kluttz. Düsseldorf: Beton-Verlag, 1984.

Documents changes in long-span building design since World War II. Includes a history of bridges prior to 1950. Historical analysis not very sophisticated.

See also: 70, 112, 113, 123, 237, 293, 294, 301, 303, 304, 310, 322, 328, 342, 344, 356, 362, 384, 593, 599, 642, 652, 656, 677, 801, 888, 893.

THE THEORY OF MACHINES AND THE KINEMATICS OF MECHANISMS

Primary Sources

993. Ampère, André-Marie. **Essai sur la philosophie des sciences, ou exposition analytique d'une classification naturelle de toutes les con-**

naissances humaines. 2 vols. Paris:
Bachelier, 1838-43.

Discusses the classification of mechanisms,
including the work of Monge and Lanz. Excludes
the study of force and considers only changes in
motions. Coins the term cinématique.

994. Banks, John. On the Power of Machines. Kendal:
Pennington, 1803.

Early work on the theory of machines. Includes a description of an instrument for measuring the air expelled from bellows, a study of
parallel motion, a description of the construction of a crank, and experiments on the strength
of wood and iron.

995. _____. A Treatise on Mills. London: W.
Richardson, 1795.

Early work on the application of mechanics to
mills. Contains parts on circular motion, moving bodies, machines and engines, the velocity
of effluent water, and a study of waterwheels.

996. Barlow, Peter. A Treatise on the Manufactures
and Machinery of Great Britain. London:
Baldwin and Cradock, 1836.

Survey of British mechanical engineering.

997. Belanger, Jean Baptiste Charles Joseph. Traité
de cinématique. Paris: Dunod, 1864.

Applies the geometrical method to the study
of mechanisms. Focuses on changes in motion.
Uses Willis' idea of relative velocities.

998. Bernoulli, Christoph. Vademecum des
Machanikers, oder Praktisches Handbuch für
Mechaniker, Machinen- und Mühlen-bauer und
Techniker überhaupt. 2 volumes. Stuttgart:
J.G. Cotta, 1829-30.

Handbook on mechanical engineering and the
theory of machines. Contains tables and formulae. Went through several editions.

999.   Betancourt y Molina, Agustin de, and Phillippe
        Louis Lanz. **Essai sur la composition des ma-
        chines.** Paris: Impr. impériale, 1808.

        Important French work on the theory of ma-
        chines. Puts forward a classification of mecha-
        nisms. Used as a textbook at the École Poly-
        technique. Translated into English as **Analyti-
        cal Essay on the Construction of Machines.**
        London: R. Ackerman, 1817.

1000.   Carnot, Lazare Nicolas Marguerite. **Essai sur
        les machines en général.** Dijon: A.-M. Defay,
        1782.

        Important early work on the theory of ma-
        chines. Applies the concept of <u>vis viva</u> to ma-
        chines. Concludes that all shocks and impacts
        must be avoided in working machines. Later edi-
        tion published as **Principes fondamentaux de
        l'équilibre et du mouvement.** Paris: Deterville,
        1803. See item 333.

1001.   Coriolis, Gaspard Gustave. **Du calcul de l'effet
        des machines, ou considérations sur l'emploi
        des moteurs et sur leur evaluation, pour
        servir d'introduction à l'étude spéciale des
        machines.** Paris: Carilian-Goeury, 1829.

        Fundamental work on the theory of machines by
        a leading French theoretician.

1002.   _____. **Mémoire sur l'influence du moment
        d'inertia du balancier d'une machine à vapeur
        et de sa vitesse moyenne sur la régulartié du
        mouvement de rotation que le va-et-vient du
        piston communique au volant.** Paris: Im-
        primerie royale, 1832.

        Theoretical study of fly-wheels and moments
        of inertia.

1003.   _____. **Traité de la mécanique des corps
        solides et du calcul de l'effet des machines.**
        Paris: Carilian-Goeury, 1844.

        Includes an important contribution to the
        theory of machines.

1004.   Coulomb, Charles Augustin de. **Théorie des ma-
        chines simples, en ayant égard au frottement
        de leurs parties et à la roideur des
        cordages.** Paris: Bachelier, 1809.

        Important work by a leading French theoreti-
        cian. Includes a discussion of friction, tor-
        sion, windmills, and mechanical movement. Based
        on papers published in the **Mémoires de
        l'Académie des sciences.**

1005.   Evans, Oliver. **The Young Mill-Wright and
        Miller's Guide.** Philadelphia, 1795.

        Surveys the principles of mechanisms and de-
        velops a set of rules for designing mills. At-
        tempts to apply scientific methods to technol-
        ogy, including the use of experiments.

1006.   Fairbairn, William. **The Principles of Mechanism
        and Machinery Transmission.** Philadelphia:
        Henry Carey Baird & Co., 1882.

        Important work on the theory of mechanism,
        especially adapted for American engineers.

1007.   _____. **Treatise on Mills and Millwork.** 2 vol-
        umes. London: Longman, Green, Longman,
        Roberts, & Green, 1861-63.

        Important work on the application of the
        principles of mechanics to the design of mills.
        Part I discusses water-wheels, turbines and
        steam engines, while part II discusses the the-
        ory of mechanisms and the strength of materials.

1008.   Girault, Charles François. **Éléments de géomé-
        trie appliquée à la transformation du mouve-
        ment dans les machines.** Caen: A. Hardel,
        1858.

        Applies geometrical methods to the study of
        mechanisms.

1009.   Goodeve, T.M. **The Elements of Mechanism.** New
        edition. London: Longmans, Green, & Co.,
        1894.

        Emphasizes actual machine parts. Written by
        a professor at the Royal School of Mines. First
        published in 1860.

1010.   Grashof, Franz. **Theoretische Maschinenlehre.** 3
         volumes. Leipzig, 1871-86.

         Fundamental textbook on the theory of ma-
         chines by one of the founders of the Verein
         Deutscher Ingenieure and the successor to
         Redtenbacher at Karlsruhe.

1011.   Gregory, Olinthus Gilbert. **A Treatise of Me-
         chanics, Theoretical, Practical, and Descrip-
         tive.** 3rd edition. London: F.C. and J.
         Rivington, 1815.

         One of the first books in English to present
         Hachette's system of mechanical movements. In-
         cludes sections on the theory of statics, dynam-
         ics, hydrostatics, hydrodynamics, and pneumat-
         ics. Also includes sections on the construction
         of machines.

1012.   Hachette, Jean Nicolas Pierre. **Traité élémen-
         taire des machines.** Paris: J. Klostermann,
         1811.

         Fundamental work on the classification of
         mechanisms. Helped to create the field of the
         kinematics of mechanisms. Focuses on changes in
         the direction of motion.

1013.   Haton de la Goupillière, Julien Napoleon. **Cours
         de machines.** 2 vols. Paris: Dunod, 1889-92.

         Course on the theory of machines by a profes-
         sor at the École Supérieure Nationale des Mines.
         Also includes discussion of steam engines and
         hydraulic machines.

1014.   _____. **Traité des mécanismes, renfermant la
         théorie géométrique des organes et celle des
         résistances passives.** Paris: Gauthier-
         Villars, 1864.

         Applies geometrical methods to the study of
         mechanisms. Classifies mechanisms according to
         the method of transmission of motion.

1015.   Hughes, William Carter. **The American Miller and
         Millwright's Assistant.** Philadelphia, 1851.

> Presents an application of the scientific method to mills.

1016. Kennedy, Alex B.W. **The Mechanics of Machinery.** London: Macmillan, 1886.

> Fundamental work on the theory of machines.

1017. Kimball, D.S. **Elements of Machine Design.** New York: J. Wiley, 1909.

> Important work on developing a science of machine design.

1018. Laboulaye, Charles Pierre Lefebvre. **Traité de cinématique: ou Théorie des mécanismes.** Paris: L. Mathias, 1849.

> Attempts a theoretical method of studying mechanisms based on Ampère (see item 993). Divides the "machine-elements" into the système levier, système tour, and système plan based on making one, two, or three points of the moving body fixed. Derives a kinematic theory from a priori principles.

1019. MacCord, Charles William. **Kinematics. A Treatise on the Modification of Motion, as Affected by the Forms and Modes of Connection of the Moving Parts of Machines.** New York: John Wiley, 1883.

> Used as a textbook to introduce mechanical engineers to the theory of kinematics. Published in several editions.

1020. McKay, Robert Ferrier. **Principles of Machine Design.** New York: Longmans, 1924.

> Develops a theory of machine design.

1021. _____. **The Theory of Machines.** London: E. Arnold, 1915.

> Contains one part on mechanics and a second part on the kinematics of machines. Applied mechanics to a theory of machines.

1022. Maxwell, James Clerk. "On Governors." **Proceedings of the Royal Society of London** 16 (1868): 270.

Fundamental work on the theory of governors.
Develops an early idea of feedback control.

1023. Moseley, Henry. "Researches in the Theory of
       Machines." **Philosophical Transactions of the
       Royal Society of London** 131 (1841): 285-305.

       Applies the concept of vis viva to machines.
       One of the first applications of the concept of
       work (one-half vis viva) to machines.

1024. Overman, Frederick. **Mechanics for the Mill-
       wright.** Philadelphia, 1864.

       Presents the application of the scientific
       method to mills.

1025. Parent, Antoine. **Éléments de mécanique et de
       physique. Ou l'on donne géométriquement les
       principles du choc & des équilibres entre
       toutes sortes de corps. Avec l'explication
       naturelle des machines fondamentales.** Paris:
       F.& P. Delaulne, 1700.

       Early theoretical analysis on the application
       of mechanics to machines.

1026. Poinsot, Louis. **Théorie des cones circulaires
       roulants.** Paris: Bachelier, 1852.

       Uses the geometrical method to describe mech-
       anisms.

1027. _____. **Théorie nouvelle de la rotation des
       corps.** Paris: Bachelier, 1834.

       Applies the geometrical method to the study
       of mechanisms. Influenced other works. Trans-
       lated into English by Charles Whitley as **A New
       Theory of the Rotation of Bodies.** Cambridge: R.
       Newby, 1834.

1028. Poncelet, Jean Victor. **Cours de mécanique ap-
       pliquée aux machines, professe en 1825-26 à
       l'École royal de l'artillerie et du génie.**
       Metz, 1828.

       Important work on the application of mechan-
       ics to a theory of machines by a leading French
       theoretician. Based on lectures at the École

Royal de l'Artillerie et du Génie. Many later
editions published.

1029. Prony, Gaspard Clair François Marie Riche, Baron
de. **Mémoir sur un moyen de convertir mouvements circulaires continus en mouvements
rectilignes alternatifs.** 3rd edition.
Paris: Bachelier, 1839.

Kinematical study of the change from circular
motion to alternating rectilinear motion by a
leading French theorist.

1030. _____. "Note sur un moyen des mesurer l'effet
dynamique des machines de rotation." **Annales
de chimie et de physique** 19 (1822): 165-73.

Describes a method for measuring the work of
a machine.

1031. Rankine, William John Macquorn. **A Manual of Machinery and Millwork.** London: Charles
Griffin & Co., 1869.

Theoretical study of machines based on applied mechanics. Considers a machine to be com-
posed of a frame and moving pieces. Shows how
mechanisms can be linked into elementary combi-
nations. Used as a textbook. Went through several editions.

1032. Redtenbacher, Ferdinand Jacob. **Die Bewegungs-
Mechanismen; Darstellung und Beschreibung
eines Theiles der Maschinen-Modell-Sammlung
der Polytechnischen Schule in Carlsruhe.**
Heidelberg: F. Bassermann, 1866.

Important work on the theory of mechanisms
and mechanical movements by the professor of mechanical engineering at the Karlsruhe
Polytechnic.

1033. _____. **Der Maschinenbau.** 3 volumes.
Mannheim: Friedrich Bassermann, 1862-65.

Fundamental work on the practical construction of machines based on a scientific theory of
machines by the Director of the Karlsruhe Polytechnic.

1034.   _____. **Prinzipien der Mechanik und des Maschinenbaues.** Mannheim: Friedrich Bassermann, 1852.

Important work on the theoretical science of machinery.

1035.   _____. **Resultate für den Maschinenbau.** Mannheim: Friedrich Bassermann, 1848.

Important work on a scientific theory of machines. Provides results without using the deductive process. Helped convince engineers of the value of theory.

1036.   Reynolds, Osborne. "On the Theory of Lubrication and Its Application to Mr. Beauchamp Tower's Experiments." **Philosophical Transactions of the Royal Society of London** 177 (1886):157-234.

Important work on the theory of lubrication by a leading engineering scientist. Applies hydrodynamical laws to the oil in bearings.

1037.   Reuleaux, Franz. **The Constructor, a Handbook of Machine Design.** Translated from the 4th German edition by Henry Harrison Suplee. Philadelphia: H.H. Suplee, 1895.

First English translation of an important contribution to machine design. Contains elements of Reuleaux's theory of machine classification, theory of kinematics, and applied kinematics. First German edition published in 1861.

1038.   _____. **The Kinematics of Machinery: Outlines of a Theory of Machines.** Translated and edited by A.B. Kennedy. London: Macmillan & Co., 1876.

Fundamental work on the theory of machines. Provided a new theory of the classification of mechanisms.

1039.   Ripper, William. **A Course of Instruction in Machine Drawing and Design for Technical Schools and Engineering Students.** London: Percival, 1890.

Widely used textbook on machine design by a professor at University College, Sheffield.

1040. Schwamb, Peter. **Notes on the Elements of Mechanism.** Boston: J.S. Cushing & Co., 1885.

Popular textbook used in M.I.T. mechanical engineering courses.

1041. Stamm, Ernest. **Essai sur l'automatique pur, suivis quelques études complémentaires d'application.** Milan: G. Daelli, 1863.

Provides a study of the realization in mechanisms of motions described by mathematical expression.

1042. Thurston, Robert H. **Friction and Lubrication, Determinations of the Laws and Coefficients of Friction by New Methods and with New Apparatus.** New York: The Railroad Gazette, 1879.

Describes experimental studies on friction.

1043. _____. **A Treatise on Friction and Lost Work in Machinery and Millwork.** New York: John Wiley & Sons, 1885.

Applied the theory of work to machines. Concludes that friction is the sole cause of lost work. Presents a systematic study of friction in machines. Went through several editions.

1044. Unwin, William Cawthorne. **The Elements of Machine Design.** London: Longmans, Green & Co., 1877.

Textbook on machine design by a professor of hydraulic and mechanical engineering at the Royal Indian Civil Engineering College. Went through several editions.

1045. Warren, S. Edward. **Elements of Machine Construction and Drawing: or, Machine Drawing, with Some Elements of Descriptive and Rational Cinematics.** New York: John Wiley & Son, 1870.

A textbook for advanced students by a professor at Rensselaer Polytechnic Institute.

1046. Willis, Robert. "Machines and Tools for Working
       in Metal, Wood, and Other Materials." **Lec-
       tures on the Results of the Great Exhibition
       of 1851.** London: David Bogue, 1852, pp. 293-
       320.

       Argues for a "more intimate union" between
       scientific and practical men. Presents a theory
       of machines based on applied mechanics.

1047. _____. **Principles of Mechanism, Designed for
       Students in Universities and for Engineering
       Students Generally.** London: J.W. Parker,
       1841.

       Provided a new generalized theory of mecha-
       nisms. Classifies mechanisms according to the
       relationship of motions rather than the applied
       motion. Karl Marx was influenced by Willis'
       theory of machines.

1048. _____. **A System of Apparatus for the Use of
       Lecturers and Experimenters in Mechanical
       Philosophy, Especially in Those Branches
       which are Connected with Mechanism.** London:
       J. Weale, 1851.

       Describes mechanical models for conducting
       experiments on the theory of mechanism.

1049. Woods, Arthur T., and Albert Stahl. **Elementary
       Mechanism: A Text-Book for Students of Me-
       chanical Engineering.** 4th edition. New
       York: D. Van Nostrand, 1893.

       Textbook on the theory of mechanism.

1050. Zeuner, Gustave Anton. **Die Schiebersteuerungen;
       mit besonderer Berücksichtigung der Steurun-
       gen bei Locomotiven.** Freiberg: J.G.
       Engelhardt, 1858.

       Study of valve-gears, slide valves and link
       motion. Translated into English as **Treatise on
       Valve Gears, with Special Consideration of the
       Link Motions of Locomotive Engines.** London:
       Spon, 1869.

See also: 388, 399, 634, 675, 685, 701, 709, 716, 790,
791.

Secondary Sources

1051.   Arnold, Gerhard.   **Bilder aus der Geschichte der
        Kraftsmaschinen, von der Antik bis zum Beginn
        des 20. Jahrhunderts.**   Munich: Moos, 1968.

        Provides a survey of the history of machines
        from antiquity to the beginning of the 20th cen-
        tury.

1052.   Bardell, P.S.   "Some Aspects of the History of
        Journal Bearings and Their Lubrication."
        History of Technology   4 (1976): 1-30.

        Describes the early friction experiments of
        Coulomb, Rennie and Morin. Discusses Osborne
        Reynolds's application of hydrodynamical laws to
        the oil in bearings. Shows how the work on lu-
        brication is tied to the development of rail-
        roads.

1053.   Bennett, Stuart.   **A History of Control Engineer-
        ing, 1800-1930.**   Sterenage, England and New
        York: Peter Peregrinus on behalf of the In-
        stitution of Electrical Engineers, 1979.

        An internalist survey of control devices from
        the time of Watt's mechanical governor to the
        development of electronic controls. Discusses
        the relationships between the study of dynamics
        and control theory. Little discussion of social
        or intellectual factors. Can be useful as a
        source book.

1054.   Bogoliubov, Alekei.   **Istoriia mekhaniki mashin.**
        Kiev: Naukova Dumka, 1964.

        Provides a history of the theory of the kine-
        matics of mechanisms.

1055.   Calder, Ritchie.   **The Evolution of the Machine.**
        Eugene Ferguson consultant.   New York: Ameri-
        can Heritage, 1968.

        Provides a history of the development of ma-
        chines.

1056.   Chapuis, Alfred, and Edmond Droz. **Automata: A Historical and Technological Study.** Translated by Alec Reid. New York: Central Book Co., 1958.

Provides a useful survey to the development of automata.

1057.   Conway, H.G. "Some Notes on the Origins of Mechanical Servo Mechanisms." **Transactions of the Newcomen Society** 29 (1953-55): 55-75.

Contains a study of the development of control devices during the 19th century.

1058.   Ferguson, Eugene S. "Kinematics of Mechanisms from the Time of Watt." **United States National Museum Bulletin**, no. 228. Washington, D.C.: Smithsonian Institution, 1962.

Provides a useful survey and insightful study of the theory of mechanisms during the 18th, 19th and 20th centuries.

1059.   Gross, Walter E. "Theory and Invention: The Case of Charles Talbot Porter and His Steam Engine Governor." **Annals of Science** 29 (1972): 257-69.

Describes Porter's contribution to the development of control theory.

1060.   Hartenberg, Richard S., and Jacques Denavit. **Kinematic Synthesis of Linkages.** New York: McGraw-Hill, 1964.

Provides a history of the theory of mechanisms and how this history contributes to modern engineering theory.

1061.   _____. "Men and Machines, an Informal History." **Machine Design** 28 (May 3, 1956): 74-82; (June 14, 1956): 101-109; (July 12, 1956): 84-93.

Surveys, with illustrations, the history of machines. Also contains a useful annotated bibliography.

1062.   Hoppe, Brigitte.   "Biologische und technische
        Bewegungslehre im 19. Jahrhundert." **Georg-
        Agricola-Gesellschaft, Geschichte der Natur-
        wissenschaften und der Technik im 19.
        Jahrhundert.** Düsseldorf: VDI-Verlag, 1969,
        pp. 9-35.

        Discusses the biological and technical roots
        of 19th-century kinematics.

1063.   Hubka, Vladimir. **Theorie der Machinensysteme;
        Grundlagen einen wissenschaftlichen Konstruk-
        tionslehre.** Berlin: Springer-Verlag, 1972.

        Provides a theory of machines using ideas
        drawn from cybernetics. Attempts to be a con-
        temporary heir to Franz Reuleaux.

1064.   Khramoi. A.V. **The History of Automation in Rus-
        sia before 1917.** Translated by the Israel
        Program for Scientific Translation.
        Jerusalem, 1969.

        Contains a history of feedback control in
        Russia.

1065.   Mayr, Otto. **Feedback Mechanisms in the Histori-
        cal Collections of the National Museum of
        History and Technology.** Washington, D.C.:
        Smithsonian Institution Press, 1971.

        Provides a study of feedback mechanisms based
        on the Smithsonian collection. Discusses the
        theory of feedback mechanisms and their use in
        steam engines, water wheels, turbines, steam
        boilers, textile machines and other devices.

1066.   _____. **The Origins of Feedback Control.**
        Cambridge: M.I.T. Press, 1970.

        Surveys the development of feedback-control
        devices from the time of the Greeks. Argues
        that a theory of feedback control did not de-
        velop until the 18th and 19th centuries. Ana-
        lyzes the contributions of James Watt and James
        Clerk Maxwell to the development of the mechani-
        cal governor. Important book.

1067.   _____. "Victorian Physicists and Speed Regula-
         tion: An Encounter between Science and Tech-
         nology." **Notes and Records of the Royal So-
         ciety of London** 26 (1971): 205-28.

         Focuses on the works of G.B. Airy, C.W.
         Siemens, L. Foucault, J.C. Maxwell, Lord Kelvin,
         and J.W. Gibbs and their contributions to speed
         regulation and automatic control. Discusses how
         science influences technology.

1068.   _____. "Yankee Practice and Engineering The-
         ory: Charles T. Porter and the Dynamics of
         the High-Speed Steam Engine." **Technology and
         Culture** 16 (1975): 570-602.

         Shows that even a practical, self-taught en-
         gineer such as Charles Porter developed an in-
         terest in the theoretical aspects of engineering
         because of intellectual curiosity concerning
         problems of the dynamics of high-speed engines.
         Includes a discussion of the development of a
         theory of the dynamics of high-speed engines.

1069.   Strandh, Sigrard. **Machines: An Illustrated His-
         tory.** London: Artists House, 1979.

         A popular history with numerous illustra-
         tions. Neglects several important people. Con-
         tains some error of interpretation and under-
         standing.

1070.   Wislicki, Alfred. "Transformability of Basic
         Models of Machines." **History and Technology**
         2 (1985): 235-244.

         Argues that the idea of archetypal machines
         provided models for the development of technol-
         ogy. Concludes that certain regularities emerge
         from a study of machines which are similar to
         the paradigms of Thomas Kuhn.

See also: 13, 246, 262, 299, 302, 319, 320, 341, 355,
369, 378, 381, 443, 446, 562, 597, 598, 638, 662, 669,
681, 801.

CHAPTER VII: THERMODYNAMICS AND HEAT TRANSFER

Primary Sources

1071. Beau de Rochas, Alphonse. **Nouvelles recherches sur les conditions pratiques de plus grand utlisation de la chaleur et en général de la fonce motrice avec application au chemin de fer et à la navigation.** Paris: E. Lacroix, 1862.

Applies the laws of thermodynamics to heat engines, including the internal combustion engine. Shows that air must be compressed in order to gain maximum efficiency. Sets forward the conditions for what would become the four stroke engine.

1072. Bernoulli, Christoph. **Handbuch der Dampf-maschinen-lehre für Teckniker und Freunde der Mechanik.** Stuttgart and Tubingen: J.G. Cotta, 1823.

Early 19th-century textbook on the steam engine.

1073. Bertrand, Joseph Louis François. **Thermo-dynamique.** Paris: Gauthier-Villars, 1887.

Important work on thermodynamics by a member of the École Polytechnique.

1074. Bourne, John. **Catechism of the Steam Engine, Illustrative of the Scientific Principles Upon Which Its Operation Depends, and the Practical Details of Its Structure, in Its Applications to Mines, Mills, Steam Navigation and Railways, with Various Suggestions of Improvement.** London: J. Williams, 1847.

Popular book on the scientific principles of the steam engine. Went through numerous editions. American edition published in 1876.

1075. _____. **Handbook of the Steam-Engine. Containing All the Rules Required for the Right Construction and Management of Engines of Every Class with the Easy Arithmetical Solutions of Those Rules.** London: Longman, Green, Longman, Roberts, & Green 1865.

Provides a key to Bourne's Catechism of the Steam Engine (item 1074). Went through several editions.

1076. _____. **Recent Improvements in the Steam Engine in Its Various Applications to Mines, Mills, Steam-Navigation, Railways, and Agriculture.** London: Longman, Green Longman, Roberts, & Green, 1865.

Provides an introduction and supplement to Bourne's Catechism (item 1074).

1077. _____, ed. **A Treatise on the Steam Engine in Its Application to Mines, Mills, Steam Navigation, and Railways.** London: Longman, Brown, Green, and Longmans, 1846.

Edited for the Artizan Club. Included sections on the theoretical investigation of the motive power of heat. Went through several editions.

1078. Carnot, Sadi Nicolas Léonard. **Réflexions sur la puissance motrice du feu et sur les machines propres à développer cette puissance.** Paris: Bachelier, 1824.

Pioneering work on the motive power of heat. Puts forward the idea of the "Carnot cycle." Helped to create the science of thermodynamics and influenced the development of engineering thermodynamics. Modern edition published as **Reflexions on the Motive Power of Fire: A Critical Edition with the Surviving Manuscripts.** Edited and Translated by Robert Fox. Manchester: Manchester University Press, 1986.

1079.   Cazin, Achille. **The Phenomena and Laws of Heat.**
        Translated and edited by Elihu Rich. New
        York: Charles Scribner & Co., 1869.

        American edition of a French work that ap-
        plies Tyndall's theory of heat as motion to ma-
        chines. First French edition published in 1867.

1080.   Clapeyron, Emile. "Mémoire sur la puissance
        motrice de la chaleur." **Journal de l'École
        Polytechnique** 14 (1834): 153-190.

        Important paper by a leading French theoreti-
        cian. Introduces the indicator diagram (or P-V
        diagram) into Sadi Carnot's theory.

1081.   Clark, D.K. "On the Expansive Working of Steam
        in Locomotives." **Proceedings of the Institu-
        tion of Mechanical Engineers** (1852): 60-88.

        Provides a study of the thermodynamics of us-
        ing the expansive power of steam in locomotives.

1082.   Clausius, Rudolf Julius Emmanuel. **Abhandlungen
        über die Mechanische Wärmetheorie.** 2 vol-
        umes. Braunschweig: F. Vieweg, 1864-67.

        Work on the mechanical theory of heat by one
        of the founders of thermodynamics. Influenced
        many works on engineering thermodynamics. Vol-
        ume 1 contains a collection of individual papers
        published by Clausius during 1850-64, including
        papers on the application of the mechanical the-
        ory of heat to the steam engine. An expanded
        three volume version was published in 1876-91
        under the title **Die Mechanische Wärmetheorie.**
        English translation by J. Hirst titled **The Me-
        chanical Theory of Heat** published in 1867 with a
        revised edition by W. Browne in 1879.

1083.   Colburn Zerah. **An Inquiry into the Nature of
        Heat, and into Its Mode of Action in the Phe-
        nomena of Combustion, Vaporisation &c.**
        London: E. & F.N. Spon, 1863.

        Study of heat by a leading engineer and edi-
        tor of the journal **Engineering.**

1084.   _____. **The Locomotive Engine; Including a De-
          scription of Its Structure, Rules for Esti-
          mating Its Capabilities and Practical Obser-
          vations on Its Construction and Management.**
          Boston: Redding & Co., 1851.

          Theoretical and practical study of the rail-
          road locomotive. Went through several editions.

1085.   _____. **Locomotive Engineering, and the Mecha-
          nism of Railways: A Treatise on the Princi-
          ples of Construction of the Locomotive En-
          gine, Railway Carriages, and Railway Plant.**
          Glasgow: Collins, 1864.

          Includes a discussion of the theory of heat
          engines as applied to railway locomotives.

1086.   _____. **Steam Boiler Explosions.** London:
          Weale, 1860.

          Study of the causes of steam boiler explo-
          sions. Slightly expanded American edition pub-
          lished in New York by D. Van Nostrand in 1873
          and an edition in 1890 with an introduction by
          Robert H. Thurston.

1087.   Cotterill, James Henry. **The Steam Engine Con-
          sidered as a Heat Engine.** London: E. & F.
          Spon, 1878.

          Based on lectures given at the Royal School
          of Naval Architecture and Marine Engineering.
          Later expanded version published in 1890 as **The
          Steam Engine Considered as a Thermodynamic Ma-
          chine.**

1088.   Desormes, C.B., and Nicolas Clément. "Mémoire
          sur la théorie des machines à feu." **Bulletin
          des sciences par la Société Philomatique des
          Paris** (1819): 115-18.

          Early paper on the theory of the steam en-
          gine. Influenced the work of Sadi Carnot.

1089.   Diesel, Rudolf. **Die Entstehung des Dieselmo-
          tors.** Berlin: J. Springer, 1913.

          Fundamental work on the Diesel engine.

1090.   _____. **Theorie und Konstruktion eines ra-
         tionellen Wärmemotors zum Erastz der Dampf-
         maschinen und der heute bekannten Verben-
         nungsmotoren.** Düsseldorf: VDI-Verlag, 1986.

         Reprint of Diesel's fundamental work on the
         theory of heat engines.

1091.   Duhamel, Jean Marie Constant. **Théorie mathéma-
         tique de la chaleur.** Paris: Guiraudet, 1834.

         Mathematical theory of heat by a foremost
         French theoretician. Influenced engineering
         thermodynamics.

1092.   _____. **Mémoire sur la méthode générale rela-
         tive au mouvement de la chaleur dans les
         corps solides plongés dans des milieux dont
         la température varie avec le temps.** Paris:
         Imprimerie royale, 1832.

         Paper on heat transfer and heat conduction in
         solids.

1093.   Dupré, Athanase. **Théorie mécanique de la
         chaleur.** Paris: Gauthier-Villars, 1869.

         Work on the mechanical theory of heat.

1094.   Farey, John. **A Treatise on the Steam Engine,
         Historical, Practical, and Descriptive.**
         London: Longman, Rees, Orme, Brown, & Green,
         1827.

         Early work on steam power. Includes a dis-
         cussion of the theory of the steam engine.

1095.   Fourier, Jean Baptiste Joseph. "Mémoire
         d'analyse sur le mouvement de la chaleur dans
         les fluides." **Mémoires de l'Académie royale
         des sciences** 2nd series 12 (1833): 507-30.

         Paper on the theory of heat transfer in flu-
         ids. Influenced engineers.

1096.   _____. **Théorie analytique de la chaleur.**
         Paris: F. Didot, 1822.

         Fundamental scientific work on the theory of
         heat. Influenced many engineers and influenced
         work on the theory of the steam engine. Trans-

lated into English as **The Analytical Theory of Heat.** Cambridge: Cambridge University Press, 1878.

1097.   _____. "Théorie du mouvement de la chaleur dans les corps solides." **Mémoires de l'Académie royale des sciences** 2nd series 4 (1824): 185-555.

          Important paper on the theory of heat transfer in solids. Influenced engineers.

1098.   French, Lester G. **Steam Turbines.** New York: McGraw-Hill, 1907.

          Authoritative work on the steam turbine.

1099.   Gibbs, J. Willard. "Graphical Methods in the Thermodynamics of Fluids." **Transactions of the Connecticut Academy** 2 (1873): 309-42.

          Fundamental paper on a graphical approach to thermodynamics. Influenced 20th-century engineering thermodynamics.

1100.   _____. "A Method of Geometrical Representation of the Thermodynamic Properties of Substances by Means of Surfaces." **Transactions of the Connecticut Academy** 2 (1873): 382-404.

          Fundamental paper which develops a geometrical approach to thermodynamics. Influenced 20th-century engineering thermodynamics.

1101.   Gray, John Macfarlane. "The Rationalization of Regnault's Experiments on Steam." **Proceedings of the Institution of Mechanical Engineers** (1889): 399-450.

          Applies Victor Regnault's experiments on steam (see item 1133) to the steam engine. Gray helped to make thermodynamics an essential tool for engineers.

1102.   Guyonneau de Pambour, François Marie. **Calcul de force des machines à vapeur, pour la navigation ou l'industrie, et pour l'achat des machines.** Paris: Bachelier, 1845.

          Early work on determining the work of steam engines. Translated into German.

1103.   _____. **Théorie de la machine à vapeur; ouvrage
          destiné à prouver l'inexactitude des méthodes
          en usage pour évalver les effets ou les pro-
          portions des machines à vapeur.** Paris:
          Bachelier, 1839.

          Examines the thermodynamics of the steam en-
          gine. Published in English as **The Theory of the
          Steam Engine.** London: J. Weale, 1839.

1104.   _____. **Traité théorique et pratique des ma-
          chines locomotives.** Paris: Bachelier, 1835.

          Important work on the thermodynamics of loco-
          motives. Translated into English as **A Practical
          Treatise on Locomotive Engines upon Railroads.**
          London: J. Weale, 1836. Also published in Amer-
          ican and German editions.

1105.   Hachette, Jean Nicolas Pierre. **Histoire des ma-
          chines à vapeur, depuis leur origine jusqu'à
          nos jours.** Paris: Corby, 1830.

          Early history of the steam engine by a lead-
          ing French theoretician and one of the founders
          of the École Polytechnique.

1106.   Hirn, Gustave Adolphe. **Mémoire sur la thermo-
          dynamique.** Paris: Gauthier-Villars, 1867.

          Study on thermodynamics by a leading theo-
          rist. Influenced the development of engineering
          thermodynamics.

1107.   _____. **Recherches sur l'équivalent mécanique
          de la chaleur, présentées à la Société de
          Physique de Berlin.** Colmar: Bureau de la Re-
          vue D'Alsace, 1858.

          Study of the mechanical equivalent of heat.

1108.   _____. **Remarques sur un principe de physique
          d'ou part M. Clausius dans sa nouvelle théo-
          rie des moteurs à vapeur.** Paris: Gauthier-
          Villars, 1888.

          Comments on Rudolf Clausius' theory of the
          steam engine (see item 1082).

1109. _____. **Théorie mécanique de la chaleur.** 3
       volumes. Paris: Gauthier-Villars, 1868-76.

       Important work on the mechanical theory of
       heat. Includes results of experimental studies
       on heat.

1110. Isherwood, Benjamin Franklin. **Engineering
       Precedents for Steam Machinery; Embracing the
       Performances of Steamships, Experiments with
       Propelling Instruments, Condensers, Boilers,
       Etc., Accompanied by Analyses of the Same.**
       New York: Baillierme, 1858.

       Experimental studies of steamship performance
       by the Chief of the Bureau of Steam Engineering
       of the U.S. Navy.

1111. _____. **Experimental Researches in Steam Engi-
       neering.** 2 volumes. Philadelphia: W.
       Hamilton, 1863-65.

       Describes important experiments conducted for
       the U.S. Navy on the performance of steam en-
       gines at sea.

1112. Kelland, Philip. **Theory of Heat.** Cambridge: J.
       and J.J. Deighton, 1837.

       A basic work on the theory of heat. Influ-
       enced engineering thermodynamics.

1113. Kelvin, William Thomson, Baron. "An Account of
       Carnot's Theory of the Motive Power of Heat,
       with Numerical Results Deduced from
       Regnault's Experiments on Steam." **Transac-
       tions of the Royal Society of Edinburgh** 16
       (1849): 541-74.

       Significant work on the theory of thermo-
       dynamics and its application to the steam en-
       gine. Based on the work of Carnot and Regnault
       (see items 1078 and 1134).

1114. _____. "On the Dynamical Theory of Heat."
       **Transactions of the Royal Society of
       Edinburgh** 20 (1853): 261-298, 475-82; 21
       (1857): 123-71.

Fundamental paper on the dynamical theory of
heat by one of the founders of the theory of
thermodynamics. Helped formalize the second law
of thermodynamics.

1115. Lamé, Gabriel. **Leçons sur la théorie analytique
de la chaleur.** Paris: Mallet-Bachelier,
1861.

Study of heat by a leading French theoretician.

1116. Lardner, Dionysius. **Popular Lectures on Science
and Art, Delivered in the Principal Cities
and Towns of the United States.** New York:
Greeley & McElrath, 1845.

Includes lectures on the steam engine by an
important "popularizer" of scientific ideas.
Helped introduce scientific ideas to mechanics.

1117. _____. **Popular Lectures on the Steam Engine,
in Which Its Construction and Operation Are
Familiarly Explained: With an Historical
Sketch of Its Invention and Progressive Improvement.** New York: E. Bliss, 1828.

Lectures on the steam engine. Helped to fa-
miliarize mechanics with the theory of steam engine theory. Later American and English editions are titled **The Steam Engine Familiarly Explained and Illustrated.**

1118. Maxwell, James Clerk. **Theory of Heat.** London:
Longmans, Green, 1871.

A notable study on the kinetic theory of
heat. Went through several editions. Influenced late 19th century engineers.

1119. Poisson, Siméon Denis. **Théorie mathématique de
la chaleur.** Paris: Bachelier, 1835.

Fundamental work on the mathematical theory
of heat and heat transfer. Influenced engineers.

1120. Rankine, William John Macquorn. **A Manual of the
Steam Engine and Other Prime Movers.** London:
Richard Griffin & Co., 1859.

Fundamental book on the application of thermodynamics to the steam engine. Used as a textbook in Europe, America and Japan well into the
20th century. First synthesis of steam engine
theory. Went through numerous editions.

1121.   _____. "On the Geometrical Representation of
the Expansive Action of Heat, and the Theory
of Thermodynamic Engines. **Philosophical
Transactions of the Royal Society of London**
144 (1854): 115-76.

One of the fundamental papers of engineering
thermodynamics.

1122.   _____. "On the Means of Realizing the Advantages of the Air-Engines." **Edinburgh New
Philosophical Journal** New series 1 (1855):
1-32.

Important paper on the thermodynamics of air
engines. Reprinted in the **Journal of the
Franklin Institute** 3rd series 30 (1855): 246,
330.

1123.   _____. "On the Mechanical Action of Heat, Especially in Gases and Vapours." **Transactions
of the Royal Society of Edinburgh** 20 (1853):
147-90.

A major influence on the application of thermodynamics to the steam engine. Helped to create the field of engineering thermodynamics.
Rankine extended the paper in the following supplements: "Notes as to the Dynamical Equivalent
of Temperature in Liquid Water, and the Specific
Heat of Atmospheric Air and Steam." **Trans. Roy.
Soc. Edin.** 20 (1853): 191-95; "On the Power and
Economy of Single-Acting Expansive Steam Engines." **Trans. Roy. Soc. Edin.** 20 (1853): 195-
204; "On the Economy of Heat in Expansive Engines." **Trans. Roy. Soc. Edin.** 20 (1853): 205-
10; "On the Zero of the Perfect Gas Thermome-
ter." **Trans. Roy. Soc. Edin.** 20 (1853): 561-3;
"On the Mechanical Action of Heat. Section VI. A
Review of the Fundamental Principles of the Me-
chanical Theory of Heat; With Remarks on the
Thermic Phenomena of Currents of Elastic Fluids,
as Illustrating Those Principles." **Trans. Roy.
Soc. Edin.** 20 (1853): 565-89; "On the Mechanical
Action of Heat. A Correction Applicable to the

Results of the Previous Reduction of the Experiments of Messers Thomson and Joule." **Proc. Roy. Soc. Edin.** 3 (1857): 223-4; "On the Mechanical Action of Heat:--Supplement to the First Six Sections, and Section Seventh." **Proc. Roy. Soc. Edin.** 3 (1857): 287-92.

1124.   _____. "On the Thermodynamic Theory of Steam-Engines with Dry Saturated Steam, and Its Application to Practice." **Philosophical Transactions of the Royal Society of London** 149 (1859): 177-92.

Important paper on engineering thermodynamics.

1125. Reech, Ferdinand (Frédéric). **Machine à air, d'un nouveau système, dèduit d'une comparaison raisonnée des systèmes de MM. Ericsson et Lemoine.** Paris: Mallet-Bachelier, 1854.

Study of the theory of air engines by a noted theoretician. Compares the work of John Ericsson and Louis Lemoine.

1126.   _____. **Mémoire sur les machines à vapeur et leur application à la navigation.** Paris: A. Bertrand, 1844.

Important work on the theory of the steam engine and its application to navigation.

1127.   _____. **Théorie des machines motrices et des effets mécaniques de la chaleur.** Paris: Lacroix, 1869.

Based on lectures given at the Sorbonne on the mechanical action of heat. Includes work on the properties of elastic fluids.

1128.   _____. **Théorie générale des effets dynamiques de la chaleur.** Paris: Mallet-Bachelier, 1854.

Work on the theory of thermodynamics that influenced the development of engineering thermodynamics.

1129. Redtenbacher, Ferdinand Jacob. **Die calorische maschine.** Mannheim: F. Bassermann, 1853.

Study of heat engines by the professor of mechanical engineering at Karlsruhe Polytechnic.

1130.   ____. **Die Gesetze des Lokomotiv-Baues.**
Mannheim: Friedrich Bassermann,1855.

Important work on the principles of locomotives.

1131.   ____. **Die luftexpansions-maschine.** Mannheim:
F. Bassermann, 1853.

Study of air engines.

1132. Regnault, Victor. "Recherches sur la dilation
des gaz." **Annales de Chimie et de Physique**
3rd series 4 (1842): 5-63; 5 (1842): 52-83.

Paper on the expansion of gases, notable for
its influence on the thermodynamics of the steam
engine. English translation in W.W. Randall,
The Expansion of Gases by Heat. New York:
American, 1902.

1133.   ____. "Recherches sur les chaleurs, spéci-
fiques des fluides élastiques." **Comptes
Rendus** 36 (1853): 676-87.

Study of the properties of steam. Experiments and data influenced work on the thermodynamics of the steam engine. English translation in the **Philosophical Magazine** 4th series 5
(1853): 473-83.

1134.   ____. "Relation des expériences pour déter-
miner les principales lois et les données
numeriques qui entrent dans le calcul des
machines à vapeur." **Mémoires de l'Académie
des sciences** 21 (1847): 1-767.

Experiments on the properties of steam. Influenced work on the thermodynamics of the steam
engine including, the work of Rankine and
Kelvin.

1135. Reid, David Boswell. **Illustrations of the Theory and Practice of Ventilation, with Remarks
on Warming, Exclusive Lighting and Communication of Sound.** London: Longman, Brown,
Green, & Longmans, 1844.

Application of the theory of heat and gases
to the problems of ventilation.

1136.   Renwick, James. **Treatise on the Steam Engine.**
New York: G. & C. & H. Carvill, 1830.

Early American work on the theory of the
steam engine.

1137.   Reynolds, Osborne. **The General Theory of
Thermo-dynamics.** London: Institution of
Civil Engineers, 1884.

Based on a series of lectures given at the
Institution of Civil Engineers by a leading engineering scientist.

1138.   _____. **Triple-Expansion Engines and Engine-
Trials.** Edited by F.E. Idell. New York: Van
Nostrand, 1890.

Study of the role of expansion in the steam
engine.

1139.   Ripper, William. **Steam.** London: Longmans,
Green, 1889.

Work on steam power by a professor of mechanical engineering at University College,
Sheffield. Later editions published as **Heat Engines.** Widely used as a textbook.

1140.   _____. **Steam-Engine Theory and Practice.**
London: Longmans, Green, 1899.

Sequel and advanced version of item 1139.
Went through several editions.

1141.   Russell, John Scott. **On the Nature, Properties,
and Application of steam and on Steam Navigation.** Edinburgh: A. and C. Black, 1841.

Based on an article written for the 7th edition of the **Encyclopaedia Britannica** by a leading naval architect. Includes an historical
sketch of steam locomotion. Emphasizes the use
of steam power on ships.

1142.   Sennett, Richard. **The Marine Steam Engine: A Treatise for the Use of Engineering Students and Officers of the Royal Navy.** London: Longmans, Green, 1882.

Widely used textbook on the thermodynamic theory of marine steam engines. Went through over fourteen editions.

1143.   Siemens, C.W. "On the Conversion of Heat into Mechanical Effect." **Proceedings of the Institution of Civil Engineers** 12 (1853): 571-600.

Paper on the motive power of heat by a leading engineering scientist. One of the first papers to bring the work of Joule to the attention of engineers.

1144.   _____. "On the Expansion of Isolated Steam, and the Total Heat of Steam." **Proceedings of the Institution of Mechanical Engineers** (1852): 131-41.

Paper on the thermodynamics of steam.

1145.   Stirling, James. "Description of Stirling's Improved Air-Engine." **Proceedings of the Institution of Civil Engineers** 4 (1845): 348-61.

Important paper on the theory of a hot air engine.

1146.   Thurston, Robert H. **The Animal as a Machine and a Prime Motor, and the Laws of Energetics.** New York: J. Wiley & Sons, 1894.

Study of force and energy based on an analysis of animal mechanics. Investigates the relationship between machines, motors, and biological processes.

1147.   _____. **Efficiency of Steam Engines and Conditions of Economy.** Philadelphia: J.S. Smith, 1882.

Studies of the behavior of steam in the steam engine cylinder. Reprinted from the **Journal of the Franklin Institute**, February, May, and June of 1882.

1148.   _____. **A Handbook of Engine and Boiler Trials and of the Indicator and Prony Brake. For Engineers and Technical Schools.** New York: J. Wiley & Sons, 1890.

Describes experimental studies of the steam engine, steam boilers, and indicators for steam engines. Went through several editions.

1149.   _____. **Heat as a Form of Energy.** New York: Houghton, Mifflin, 1890.

Describes the scientific idea of heat and its application in gas and steam engines.

1150.   _____. **A History of the Growth of the Steam-Engine.** New York: D. Appleton & Co., 1878.

Fundamental work on the development of the steam engine by one of the leading engineering scientists in America. Went through many editions and was translated into French.

1151.   _____. **A Manual of Steam-Boilers: Their Design, Construction, and Operation.** New York: J. Wiley & Sons, 1888.

Study of steam boilers, based on experimental and theoretical studies. Used as a text in many engineering courses. Went through several editions.

1152.   _____. **A Manual of the Steam-Engine: For Engineers and Technical Schools.** 2 volumes. New York: John Wiley & Sons, 1891.

Significant work on the application of thermodynamics to the steam engine. Influenced by the work of Rankine (item 1120).

1153.   _____. **Steam Boiler Explosions, in Theory and Practice.** New York: J. Wiley & Sons, 1887.

Fundamental work on steam boiler explosions.

1154. Tredgold, Thomas. **Illustrations of Steam Machinery and Steam Naval Architecture.** London: J. Weale, 1839.

Provides supplemental material to the author's book on the steam engine (item 1156).
Discusses the application of the steam engine as
a motive power for ships.

1155.   _____. **The Principles and Practice and Explanation of the Machinery of Locomotive Engines
in Operation on the Several Lines of Railway.**
London: John Weale, 1850-53.

A new and expanded version of Tredgold's work
on the steam engine (item 1156).

1156.   _____. **The Steam Engine, Comprising an Account
of Its Invention and Progressive Improvement;
with an Investigation of Its Principles, and
the Proportions of Its Parts for Efficiency
and Strength; Detailing also Its Application
to Navigation, Mining, Impelling Machines,
&c.** London: J. Taylor, 1827.

Important early work on the principles of the
actions of the steam engine. Enlarged to three
volumes by W.S.B. Woolhouse in the 1838 edition.
Went through several editions and was translated
into French.

1157.   Wood, De Volson. **Thermodynamics.** New York:
Burr Printing House, 1887.

Textbook on thermodynamics by an important
American engineer. Went through several editions.

1158.   Zeuner, Gustav Anton. **Das Locomotiven-Blasrohr.
Experimentelle und theoretische Üntersuchun-
gen über die Zugerzeugung durch Dampf-
strahlen und über die saugende Wirkung der
Flüssigkeitsstrahlen überhaupt.** Zurich:
Meyer & Zeller, 1863.

Theoretical and experimental study of locomotives.

1159.   _____. **Grundzüge der mechanischen Wärmetheo-
rie. Mit besonderer Rücksicht auf das ver-
halten des Wasserdampfes.** Freiberg: J.G.
Engelhardt, 1860.

Fundamental work on the application of a mechanical theory of heat to the steam engine.
Published into English as **Technical Thermodynamics**. New York: D. Van Nostrand, 1906.

See also: 393, 397, 400, 401, 403, 404, 405, 410, 412,
608, 713, 745, 752, 759, 780, 790, 793, 1013, 1248,
1278, 1328, 1329, 1406.

Secondary Works

1160.   Akimov, Pavel Petrovich. **Istoriia Razvitiia
        Sudovykh Energeticheskikn Ustanovok.**
        Leningrad: Sudostroenie, 1966.

        Discusses the development of the marine steam
        engine.

1161.   Alefeld, George. "Einstein as Inventor."
        Physics Today 33 (May 1980): 9-13.

        Discusses the invention of a heat pump by
        Albert Einstein and Leo Szilard.

1162.   Angrist, Stanley W. "Perpetual Motion Ma-
        chines." **Scientific American** 218 (January
        1968): 115-22.

        Discusses the relationship between machines
        and the development of the first and second laws
        of thermodynamics.

1163.   Bachelard, Gaston. **Étude sur l'evolution d'un
        problème de physique; La propagation ther-
        mique dans les solides.** 2nd edition. Paris:
        Vrin, 1973.

        Discusses the history of heat transfer in
        solids.

1164.   Barnett, Martin. "Sadi Carnot and the Second
        Law of Thermodynamics." **Osiris** 13 (1958):
        327-57.

        Discusses the role of Carnot in the creation
        of the second law of thermodynamics.

1165.   Barton, D.B. **The Cornish Beam Engine.** Truro,
        Cornell: Barton, 1966.

Surveys the development of the Cornish beam
engine from 1800 to the present. Discusses the
relationship between increases in efficiency and
scientific knowledge of thermodynamics.

1166.   Bradley, Derek. "Thermodynamics--A Daughter of
        Steam." **Chartered Mechanical Engineer** 10
        (December 1963): 600-605.

Brief survey of the role of the steam engine
in the development of thermodynamics from Watt
to Rankine.

1167.   Breger, Herbert. **Die Natur als arbeitende Mas-
        chine: Zür Entstehung des Energiebegriffs in
        der Physik, 1840-1850.** Frankfurt: Campus
        Verlag, 1982.

Discusses the relationship between the discovery of energy and the view of nature as a la-
bor machine. Focuses on the theories of Karl
Marx.

1168.   Bryant, Lynwood. "The Development of the Diesel
        Engine." **Technology and Culture** 17 (1976):
        432-46.

Traces the role of invention, development,
and innovation in the evolution of the Diesel
engine. Discusses the role of thermodynamics in
the invention of the engine. Argues that the
processes of invention, development and innovation take place at the same time during the evolution of a technological device.

1169.   _____. "The Origin of the Automobile Engine."
        Scientific American 216 (March 1967): 102-
        12.

Discusses the role of Nicolaus Otto and his
invention of the four-stroke cycle.

1170.   _____. "The Origin of the Four-Stroke Cycle."
        Technology and Culture 8 (1967): 178-98.

Discusses the priority debate over the invention of the four-stroke cycle. Concludes it was
first successfully used by N. Otto in 1876. Describes the theory that led Otto to the invention.

1171.   _____. "The Role of Thermodynamics in the Evo-
         lution of Heat Engines." **Technology and Cul-
         ture** 14 (1973): 152-65.

         Discusses the relationship between the rise
         of thermodynamics in the first half of the 19th
         century and the rapid development of engines in
         the second half of the century. Argues that
         successful new forms of heat engines depended on
         engineers being able to grasp the concept of the
         convertibility of heat and work as well as the
         concept that only a fraction of the heat could
         be converted into work. Concludes that such
         concepts could not be gained from common sense
         or experience but had to be explicitly learned
         from science. Important paper.

1172.   _____. "Rudolf Diesel and His Rational En-
         gine." **Scientific American** 221 (February
         1969): 108-117.

         Discusses the role of thermodynamics in the
         development of the Diesel engine.

1173.   _____. "The Silent Otto." **Technology and Cul-
         ture** 7 (1966): 184-200.

         Discusses the development of the four-cycle
         engine by Nicolaus Otto. Includes a discussion
         of the role played by Franz Reuleaux. Concludes
         that an incorrect theory led Otto to a success-
         ful engine.

1174.   Cardwell, Donald S.L. **From Watt to Clausius: The
         Rise of Thermodynamics in the Early Indus-
         trial Age.** Ithaca: Cornell University Press,
         1971.

         Provides a particularly important study of
         the rise of thermodynamics and its relationship
         to water and steam power technology. Argues
         that the development of power technology influ-
         enced the thermodynamic concepts put forward by
         Carnot. Investigates the role played by the
         Cornish pumping engines in the development of
         concepts of efficiency. Important work.

1175.   _____. "Power Technologies and the Advance of
         Science, 1700-1825." **Technology and Culture**
         6 (1965): 188-207.

Argues that the development of water power
and steam engines during the 19th century contributed to Sadi Carnot's theory of the motive
power of heat. Suggests that the parallel between water and steam power played an important
role in Carnot's thinking.

1176.   _____. "Science and Technology: The Work of
           James Prescott Joule." **Technology and Cul-
           ture** 17 (1976): 674-86.

Argues that Joule's work on the dynamical
theory of heat and the conservation of energy
grew out of an attempt to develop electric motors that could be powered by batteries. Concludes that Joule's engineering work with the
heat generated from electric currents led him to
the scientific study of heat and its equivalent
in mechanical work.

1177.   _____. "Science and the Steam Engine, 1790-
           1825." **Science and Society, 1600-1900.**
           Edited by Peter Mathias. Cambridge:
           Cambridge University Press, 1972, pp. 81-96.

Attempts to support the idea that in the case
of the steam engine an advance in technology de-
termined the progress of science.

1178.   _____. "Science and the Steam Engine in the
           Early 19th Century Reconsidered." **Transac-
           tions of the Newcomen Society** 49 (1977-78):
           111-120.

1179.   _____. "Science, Technology, and Industry."
           **The Ferment of Knowledge: Studies in the His-
           toriography of Eighteenth Century Science.**
           Edited by G.S. Rousseau and Roy Porter.
           Cambridge: Cambridge University Press, 1980.

Focuses on the development of the steam engine and thermodynamics.

1180.   _____. "Some Factors in the Early Development
           of the Concepts of Power, Work and Energy."
           British Journal for the History of Science 3
           (June 1967): 209-24.

Argues that 18th-century power technology influenced the theoretical development of the
thermodynamic concepts of power, work and energy.

1181.   _____. **Steam Power in the Eighteenth Century:
         A Case Study in the Application of Science.**
         London: Sheed and Ward, 1963.

Analyzes the Newcomen and Watt engines in
terms of 18th-century scientific theories. Discusses the significance of the confusion between
heat conduction and heat capacity.

1182.   _____, and Richard L. Hills. "Thermodynamics
         and Practical Engineering in the Nineteenth
         Century." **History of Technology** 1 (1976):
         1-20.

Investigates the relationship between the
science of thermodynamics and technological development. Concludes that practical technology
played an important role in the development of
thermodynamics.

1183.   Channell, David F. "The Problem of Two Back-
         Pressures: The Development of Watt's Separate
         Condenser." **American Journal of Physics** 42
         (1974): 65-67.

Discusses the scientific background of Watt's
invention of the separate condenser. Argues
that the Newcomen engine had two back-pressures
which reduced efficiency. Concludes that in order to overcome those two back-pressures, Watt
needed a second cylinder, or separate condenser.

1184.   Clarke, J.F. **An Almost Unknown Great Man:
         Charles Parsons and the Significance of the
         Patents of 1884.** Newcastle upon Tyne:
         Newcastle upon Tyne Polytechnic, 1984.

Discusses the multistage reaction steam
turbine.

1185.   Compton, William David. "Internal Combustion
         Engines and Their Fuel: A Preliminary Explo-
         ration of Technological Interplay." **History
         of Technology** 7 (1982): 23-36.

Investigates the relationship between fuel
requirements and the design of internal combustion engines.

1186. Cummins, C. Lyle, Jr. **Internal Fire.** Lake
       Oswego, Oregon: Carnot Press, 1976.

Presents a history of the internal combustion
engine from its beginnings to 1900. Focuses on
the technology rather than history. Provides a
detailed description of engines. The first half
of the book includes a history of steam and hot
air engines before 1860. The second half of the
book focuses on the internal combustion engines
of Otto and Langen.

1187. Daub, Edward E. "Entropy and Dissipation."
       Historical Studies in the Physical Sciences
       2 (1970): 321-54.

Examines the debate between Tait and Clausius
over the discovery of the second law of thermodynamics. Concludes that Clausius was the first
to show the relationship between lost work and
irreversible entropy increases.

1188. _____. "The Regenerator Principle in the
       Stirling and Ericsson Hot Air Engines."
       British Journal for the History of Science 7
       (1974): 259-77.

1189. _____. "Sources for Clausius' Entropy Concept:
       Reech and Rankine." **Human Implications of
       Scientific Advance: Proceedings of the XVth
       International Congress of the History of Sci-
       ence.** Edited by E.G. Forbes. Edinburgh:
       University of Edinburgh Press, 1978, pp. 342-
       57.

Argues that the engineering thermodynamics of
W.J.M. Rankine and F. Reech played a significant
role in the development of Clausius' thermodynamics, especially his concept of entropy.
Concludes that Clausius was led to his concept
of entropy in order to refute Rankine's sugges-
tion that energy might be concentrating in some
other parts of the universe.

1190. Dickinson, Henry W. **A Short History of the
       Steam Engine.** London: Cass, 1963.

Reprint of the 1938 edition with a new introduction by A.E. Musson. Provides a survey of
the development of the steam engine. Includes
some discussion of the role of science.

1191.   _____, and Rhys Jenkins. **James Watt and the
Steam Engine.** Oxford: Clarendon Press, 1927.

Detailed study of Watt's improvements of the
steam engine. Based on the papers of Watt and
Matthew Boulton.

1192.   Donaldson, Robert, ed. **Bicentenary of the James
Watt Patent for a Separate Condenser for the
Steam Engine: A Symposium.** Glasgow: University of Glasgow for the James Watt Bicentenary Committee, 1971.

Contains two papers on the history of the
steam engine, including one on the historical
development of the separate condenser. The rest
of the articles are on modern engineering.

1193.   Donovan, Arthur L. "Toward a Social History of
Technological Ideas: Joseph Black, James
Watt, and the Separate Condenser." **The His-
tory and Philosophy of Technology.** Edited by
George Bugliarello and Dean B. Doner.
Urbana: University of Illinois Press, 1979,
pp. 19-30.

Focuses on the relationship between Watt and
Black and how it contributed to the invention of
the separate condenser. Includes a discussion
of the role of science.

1194.   Duffy, Michael C. "Mechanics, Thermodynamics
and Locomotive Design: Analytical Methods and
the Classical Form." **History and Technology**
2 (1985): 17-61.

Analyzes the role of mechanics and thermodynamics in the design of locomotives.

1195.   _____. "Mechanics, Thermodynamics, and Locomotive Design: The High Pressure Steam Locomotive." **History and Technology** 3 (1987): 155-
92.

Focuses on the role of thermodynamics in the development of the high pressure locomotive engine.

1196.   _____. "Mechanics, Thermodynamics, and Locomotive Design: The Machine-Ensemble and the Development of Industrial Thermodynamics." **History and Technology** 1 (1983): 45-78.

Analyzes the role of locomotive design in the development of engineering thermodynamics.

1197.   _____. "Mechanics, Thermodynamics, and Locomotive Design: The Turbine-Condenser Locomotive." **History and Technology** 3 (1987): 87-122.

Analyzes the role of thermodynamics in the development of the turbine-condenser locomotive.

1198.   Ferguson, Eugene S. "John Ericsson and the Age of Caloric." **Contributions from the Museum of History and Technology, U.S. National Museum Bulletin, 228.** Washington, D.C., 1961, pp. 41-60.

Shows that failure of Ericsson's caloric engine was the result of his application of the then current material theory of heat to technology. Concludes that Ericsson was violating the second law of thermodynamics which was just being formulated.

1199.   _____. "The Origins of the Steam Engine." **Scientific American** 210 (January 1964): 98-107.

Discusses the history of the steam engine. Includes a discussion of the role of science.

1200.   Finch, James Kip. "Tredgold: The Steam Engine." **Consulting Engineer** 21 (September 1963): 94-101.

Analyzes Tredgold's 1827 book on the steam engine (see item 1156).

1201.   Fleming, Donald. "Latent Heat and the Invention of the Watt Engine." **Isis** 43 (1952): 3-5.

Discusses the role of the scientific theory
of latent heat in the development of Watt's
steam engine.

1202. Fox, Robert. **The Caloric Theory of Gases from
       Lavoisier to Regnault.** Oxford: Clarendon
       Press, 1971.

       Discusses the history of the caloric theory
       of gases which had a major impact on 19th-cen-
       tury engineering thermodynamics.

1203. _____. "The Fire Piston and Its Origins in Eu-
       rope." **Technology and Culture** 10 (1969):
       355-70.

       Demonstrates that the fire piston, a cylinder
       for compressing air and heating it to create a
       fire, was invented in Europe some fifty years
       before it was established that work could be
       changed into heat. Discusses the role of 18th-
       century theories of heat.

1204. _____. "The Intellectual Environment of Sadi
       Carnot: A New Look." **Actes XIIe Congrès In-
       ternational d'Histoire des Sciences** 4
       (1968): 67-72.

       Focuses on the role of Charles Bernard
       Desormes and Nicolas Clément on the thermodynam-
       ics of Sadi Carnot.

1205. _____. "Watt's Expansive Principle in the Work
       of Sadi Carnot and Nicolas Clément." **Notes
       and Records of the Royal Society of London**
       24 (1970): 233-53.

       Shows how Watt's idea of the benefits of ex-
       pansion in the steam engine influenced Carnot's
       work and led him to incorporate adiabatic expan-
       sion in his cycle of an ideal heat engine. Also
       shows that Clément's theoretical calculation on
       the steam engine became a source for Carnot's
       work.

1206. Garber, Elizabeth. "James Clerk Maxwell and
       Thermodynamics." **American Journal of Physics**
       37 (1969): 146-55.

Discusses the role of Maxwell and his kinetic
theory of heat in the development of thermodynamics.

1207. Harris, F.R. "The Parsons Centenary--A Hundred
       Years of Steam Turbines." **Proceedings of the
       Institution of Mechanical Engineers** Part A,
       198 (1984): 183-224.

       Surveys the development of the reaction steam
       turbine from the time of Charles Parsons.

1208. Hiebert, Erwin N. **Historical Roots of the Con-
       servation of Energy.** Madison: State Histori-
       cal Society of Wisconsin, 1962.

       Classic study of the history of a key element
       of the theory of thermodynamics. Includes a
       discussion of the role of technology in the de-
       velopment of the conservation of energy.

1209. Hills, R.L., and A.J. Pacey. "The Measurement
       of Power in Early Steam-Driven Textile
       Mills." **Technology and Culture** 13 (1972):
       25-43.

       Argues that the application of rotary steam
       engines to the textile industry led to the theo-
       retical question of the measurement of the
       "force" needed to work individual textile ma-
       chines. Shows that the economic demand that en-
       gines had to be matched to the loads of the ma-
       chines led to a search for ways to measure power
       in rotative steam engines. Argues that such
       problems led to attempts to measure the power of
       engines as they drove machines. Traces the evo-
       lution of continuous measurement of pressure in
       the steam engine cylinder through an indicator
       diagram.

1210. Hoyer, Ulrich. "Carnots **Réflexions:** Zür
       Entstenhung der Thermodynamik vor 150
       Jahren." **Physikalische Blätter** 30 (1974):
       385-93.

1211. _____. "Considerations of Carnot's Mechanical
       Equivalent of Heat." **Human Implications of
       Scientific Advance: Proceedings of the XVth
       International Congress of the History of Sci-
       ence.** Edited by E.G. Forbes. Edinburgh:

University of Edinburgh Press, 1978, pp. 359-
67.

Discusses how Sadi Carnot came to calculate a
mechanical equivalent of heat and how it differed from the method used by Robert Mayer.

1212.   _____. "Das Verhältnis der Carnotschen Theorie
zür klassischen Thermodynamik." **Archive for
History of Exact Sciences** 15 (1976): 149-97.

Detailed mathematical analysis of Carnot's
thermodynamics including his calculation of the
mechanical equivalent of heat.

1213.   _____. "How Did Carnot Calculate the Mechanical Equivalent of Heat?" **Centaurus** 19
(1975): 207-19.

Discusses Carnot's method for calculating the
mechanical equivalent of heat.

1214.   _____. "Über den Zusammenhang der Carnotschen
Theorie mit der Thermodynamik." **Archives for
History of Exact Sciences** 13 (1974): 359-75.

Discusses the connection between Carnot's
theory of heat and the development of thermodynamics.

1215.   Hunter, Louis C. **A History of Industrial Power
in the United States, 1780-1930. Volume II:
Steam Power.** Charlottesville: University
Press of Virginia for the Hagley Museum and
Library, 1985.

Provides an analysis of the stationary steam
engine in America from 1750 to 1900. Gives only
a brief discussion of the relationship between
science and technology in the steam engine.
Suggests that most advances were practical.

1216.   Hutchison, Keith. "Der Ursprung der En-
tropiefunktion bei Rankine und Clausius."
Annals of Science 30 (1973): 341-64.

Discusses the development of the concept of
entropy in the work of Rankine and Clausius.
Argues that Rankine discovered the existence of
entropy and Clausius discovered the law of the
increase of entropy.

1217.   _____. "W.J.M. Rankine and the Entropy Func-
         tion." **Proceedings of the XIVth Interna-
         tional Congress of the History of Science** 2
         (1974): 281-84.

         Discusses the role of Rankine in the develop-
         ment of the concept of entropy. Disagrees with
         the claim that Rankine's idea is equivalent to
         conventional thermodynamics. Argues that
         Rankine's idea fails to imply irreversibility.

1218.   _____. "W.J.M. Rankine and the Rise of Thermo-
         dynamics." D.Phil. Thesis. University of
         Oxford, 1976.

         Discusses the role of Rankine in the develop-
         ment of thermodynamics. Criticizes the physical
         basis of Rankine's theory as inadequate. In-
         cludes a discussion of the role of Rankine's
         thermodynamics in power engineering and the
         steam engine.

1219.   _____. "W.J.M. Rankine and the Rise of Thermo-
         dynamics." **British Journal for the History
         of Science** 14 (1981): 1-26.

         Discusses the role of Rankine in the creation
         of thermodynamics. Emphasizes his contributions
         to science rather than to engineering thermo-
         dynamics.

1220.   Kerker, Milton. "Science and the Steam Engine."
         Technology and Culture 2 (1961): 381-90.

         Argues that the steam engine was not a purely
         craft invention which had science applied to it
         only in the 19th century. Concludes that Thomas
         Savery, Thomas Newcomen, and James Watt all used
         some scientific concepts. Argues that the sci-
         ence of pneumatics and mechanics provided a con-
         ceptual basis and the practical guidance for the
         development of the steam engine.

1221.   Klein, Martin J. "Carnot's Contribution to
         Thermodynamics." **Physics Today** 27 (1974):
         23-28.

         Discusses the role of Carnot's theory in the
         development of thermodynamics.

1222.   _____ . "The Early Papers of J. Willard Gibbs:
         A Transformation of Thermodynamics." **Human
         Implications of Scientific Advance: Proceed-
         ings of the XVth International Congress of
         the History of Science.** Edited by E.G.
         Forbes. Edinburgh: University of Edinburgh
         Press, 1978, pp. 330-41.

         Discusses the role of Gibbs in the develop-
         ment of thermodynamics. Analyzes Gibbs' use of
         geometrical methods in thermodynamics.

1223.   Kranakis, Eda Fowlks. "Blast Pipes, Injectors,
         and Theories of Heat." **Technology and Sci-
         ence: Important Distinctions for Liberal
         Arts Colleges** (item 12), pp. 38-51.

         Shows how Henri Giffard's steam injector
         caused problems in the prevailing theory of heat
         and led scientists to study the theory of the
         flow of steam. Argues that these studies led to
         the invention of the steam turbine. Concludes
         that technology can play as important a role in
         the development of scientific ideas as science
         plays in the development of technological ideas.

1224.   _____ . "The French Connection: Giffard's In-
         jector and the Nature of Heat." **Technology
         and Culture** 23 (1982): 3-38.

         Argues that technological devices which could
         not be explained by the old caloric theory of
         heat, such as Giffard's injector, were important
         stimuli in the diffusion of scientific ideas to
         engineers. Shows that such devices encouraged
         the development of academic training for engi-
         neers. Also argues that the theoretical prob-
         lems raised by such devices could not always be
         solved by scientific theories. Important case
         study.

1225.   Kuhn, Thomas S. "The Caloric Theory of Adia-
         batic Compression." **Isis** 49 (1958): 132-
         140.

         Discusses the role of Sadi Carnot in the de-
         velopment of a theory of adiabatic compression.

1226.   _____ . "Carnot's Version of 'Carnot's Cycle.'"
         American Journal of Physics 23 (1955): 91-
         95.

        Shows that in the Carnot cycle caloric was a
material fluid that was conserved; also that
quantity of heat and quantity of caloric were
interchangeable.

1227.   _____. "Engineering Precedent for the Work of
      Sadi Carnot." **Archives Internationale
      d'Histoire des Sciences** 13 (1960): 251-55.

        Describes the role of engineering in the de-
velopment of Carnot's theory of thermodynamics.

1228.   _____. "Sadi Carnot and the Cagnard Engine."
      Isis 52 (1961): 567-74.

        Describes Carnot's theory of heat engines and
its application to the buoyancy engine.

1228a. Layton, Edwin T., Jr., and John Lienhard, eds.
      **History of Heat Transfer, Essays in Honor of
      the 50th Anniversary of the ASME Heat Trans-
      fer Division.** New York: American Society of
Mechanical Engineers, 1988.

        Historical anthology on an important engi-
neering science. One of the few works on the
field.

1229. Lervig, Philip. "On the Structure of Carnot's
      Theory of Heat." **Archives for History of Ex-
      act Sciences** 9 (1972): 222-39.

        Analyzes the role of Carnot's theory of heat
in his development of thermodynamics.

1230.   _____. "Sadi Carnot and Nicolas Clément." **Hu-
      man Implications of Scientific Advance: Pro-
      ceedings of the XVth International Congress
      of the History of Science.** Edited by E.G.
Forbes. Edinburgh: University of Edinburgh
Press, 1978, pp. 293-304.

        Discusses the relationship between the work
of Clément and that of Carnot. Concludes that
Clément's theoretical calculations on the steam
engine were a source for Carnot's work.

1231.   _____. "Sadi Carnot and the Steam Engine:
         Nicolas Clément's Lectures on Industrial
         Chemistry." **British Journal for the History
         of Science** 18 (1985): 147-96.

         Discusses the relationship between Clément's
         manuscript on industrial chemistry and Carnot's
         Reflections on the Motive Power of Heat. In-
         cludes a copy of notes from Clément's course.
         Concludes that Clément's ideas helped Carnot de-
         velop his theory.

1232.   Lindqvist, Svante. **Technology on Trial: The
         Introduction of Steam Power into Sweden,
         1715-1736.** Stockholm: Almqvist & Wiksell,
         1984.

         Presents an analysis of the introduction of a
         Newcomen steam engine at the Dannemora mines in
         Sweden. Argues that Marten Triewald, a gentle-
         man scientist, helped transfer the engine from
         England to Sweden in order to prove the utili-
         tarian nature of science. Shows that a shared
         set of values and ideology was an important ele-
         ment in the relationship between science and
         technology.

1233.   Low, R.J. **James Watt and the Separate Con-
         denser: An Account of the Invention.** Science
         Museum Monograph, no. 3. New York: British
         Information Services, 1969.

         Provides a detailed technical analysis, using
         museum artifacts, of Watt's separate condenser.
         Discusses the steps that led to the invention,
         including Watt's scientific experiments.

1234.   McKie, Douglas, and Neil H. Heathcote. **The
         Discovery of Specific and Latent Heats.**
         London: Arnold, 1935.

         Traces the history of the discovery of latent
         and specific heats, especially the work of
         Joseph Black. Discusses the relationship be-
         tween Black's work and James Watt's improvement
         of the steam engine.

1235.   Marland, E.A. "Nineteenth Century Work on Heat
         Exchange and Its Interaction with Technol-
         ogy." Ph.D. dissertation. University Col-
         lege, London, 1966.

1236. Mauel, Kurt. **Die Rivalitat zwischen Heissluft-maschine und Verbrennungsmotor als Kleingewerbemaschinen zwischen 1860 und 1890. Der Sieg des Verbrennungsmotor und seine Grunde.** Düsseldorf: VDI-Verlag, 1967.

       Discusses the rivalry between the hot air engine and the internal combustion engine as a source of industrial power. Includes a discussion of the theoretical problems raised by the hot air engine.

1237. Mendoza, E. "Contributions to the Study of Sadi Carnot and His Work." **Archives Internationale d'Histoire des Sciences** 12 (1959): 377-96.

1238. Monteil, C. "Auguste Rateau, la thermodynamique et les machines motrices." **Mémoires et Comptes Rendus de la Société des Ingénieurs Civils de France** (1930): 912-16.

       Discusses Rateau's study of the thermodynamics of the heat engine.

1239. Moyer, Donald Franklin. "Energy, Dynamics, Hidden Machinery: Rankine, Thomson, Tait, and Maxwell." **Studies in the History and Philosophy of Science** 8 (1977): 251-68.

       Discusses the mechanical models used in 19th-century theories of energy.

1240. Otani, Ryoichi. "Relation between Science and Technique in the Development of the Steam Engine." **Journal of History of Science: Kagaku-shi Kenkyu** 25 (1953): 23-28.

       Article in Japanese.

1241. Pacey, A.J. "Some Early Heat Engine Concepts and the Conservation of Heat." **British Journal for the History of Science** 7 (1974): 135-45.

       Investigates the relationship between the idea of the conservation of heat and the development of heat engines.

1242.   Payen, Jacques. "La préhistoire de la théorie
        expérimentale de la machine à vapeur en
        France, 1841-1843." **Proceedings of the XIVth
        International Congress of the History of Sci-
        ence** 3 (1974): 185-88.

        Discusses the early development of a theory
        of the steam engine in France.

1243.   _____. "Une publication de 1784 en langue
        française sur la machine à vapeur à con-
        denseur." **Actes XIIe Congrès International
        d'Histoire des Sciences** 10b (1968): 63-69.

        Focuses on Demaillard's book **Théorie des ma-
        chines mues par la force de la vapeur d'eau.**

1244.   Pursell, Carroll W., Jr. **Early Stationary Steam
        Engines in America: A Study in the Migration
        of a Technology.** Washington, D.C.: Smithso-
        nian Institution Press, 1969.

        Useful study of the steam engine in America.
        Analyzes the development of steam engine design.
        Finds that American mechanics and engineers had
        contacts with British engineers and mechanics.

1245.   Raman, V.V. "Evolution of the Second Law of
        Thermodynamics." **Journal of Chemical Educa-
        tion** 47 (1970): 331-37.

        Discusses the role of Sadi Carnot in the de-
        velopment of the second law of thermodynamics.

1246.   Redondi, Pietro. **L'Accueil des idées de Sadi
        Carnot: Et la technologie française de 1820 à
        1860--de la legende à l'histoire.** Paris:
        Libairie Philosophique, 1980.

        Investigates the origins and reception in
        France of Carnot's famous work on thermodynamics
        (item 128). Distinguishes between the tradition
        of the steam engine and the tradition of the hot
        air engine. Argues that Carnot's theory in
        France was developed by those investigating the
        hot air engine, although most rejected his mate-
        rial theory of heat.

1247. _____. "Sadi Carnot et la récherche tech-
       nologique en France de 1825 à 1850: Présenta-
       tion d'un travail de récherche." **Revue
       d'Histoire des Sciences et leurs Applications**
       29 (1976): 242-59.

       Investigates the relationship between techno-
       logical developments in France and Carnot's the-
       ory of the motive power of heat.

1248. Robinson, Eric, and A.E. Musson, eds. **James
       Watt and the Steam Revolution: A Documentary
       History**. New York: Kelley, 1969.

       Contains forty-eight documents relating to
       Watt and his improvement of the steam engine.
       Includes information on Watt's use of science in
       developing the separate condenser.

1249. Sass, Friedrich. **Geschichte des deutschen Ver-
       brennungsmotorenbaues**. Berlin: Springer
       Verlag, 1962.

       Useful survey of the development of the in-
       ternal combustion engine in Germany from 1860 to
       1920.

1250. Scaife, W. Garrett. "The Parsons Steam Tur-
       bine." **Scientific American** 252 (April
       1985): 132-39.

       Discusses the role of scientific ideas in the
       invention of the steam turbine.

1251. Sheriff, Thomas. "The Early Development of the
       Ljungstrom Radial Flow Steam Turbine."
       Transactions of the Newcomen Society 52
       (1980-81): 31-47.

       Discusses the development of the steam tur-
       bine.

1252. Smeaton, William A. "Some Comments on James
       Watt's Published Account of His Work on Steam
       and Steam Engines." **Notes and Records of the
       Royal Society of London** 26 (1971): 35-42.

       Discusses Watt's scientific studies of steam
       and their relationship to his improvement of the
       steam engine.

1253.   Smith, Crosbie W.   "William Thomson and the Cre-
         ation of Thermodynamics: 1840-1855."
         **Archives for the History of the Exact Sci-
         ences** 16 (1977): 231-88.

         Discusses Thomson's (Lord Kelvin's) role in
         the creation of thermodynamics.

1254.   Sonnenberg, Gerhard Siegfried. **Hundert Jahre
         Sicherheit: Beiträge zür technischen und ad-
         ministrativen Entwicklung des Dampfkessel-
         wessens in Deutschland 1810 bis 1910.**
         Düsseldorf: VDI-Verlag, 1968.

         Describes the evolution of understanding and
         control of steam boiler explosions in Germany
         between 1810 and 1910. Provides a good case
         study on the interaction of science and technol-
         ogy and its role in government. Describes the
         scientific investigations into the cause of
         boiler explosions.

1255.   Storr, F.   "The Development of the Marine Com-
         pound Engine, c. 1870-1890."   Ph.D. disserta-
         tion. Newcastle Polytechnic, 1982.

1256.   Stowers, Arthur. "Thomas Newcomen's First Steam
         Engine 250 Years Ago and the Initial Develop-
         ment of Steam Power." **Transactions of the
         Newcomen Society** 34 (1961-62): 133-50.

         Analyzes the scientific and technological
         ideas from 1606 to 1712 which led to the steam
         engine.

1257.   Takayama, Susumu. "Development of 'the Theory
         of the Steam Engine.'" **Japanese Studies in
         the History of Science** 18 (1979): 101-116.

         Article in Japanese.

1258.   Truesdell, Clifford. **The Tragicomical History
         of Thermodynamics, 1822-1854.** New York:
         Springer, 1980.

         Internalist history of thermodynamics. Em-
         phasizes thermodynamics as a science but in-
         cludes discussion of the work of W.J.M. Rankine
         and F. Reech. Argues that Reech's role has not
         been fully appreciated.

1259.   _____, and S. Bharatha. **The Concepts and Logic
         of Classical Thermodynamics as a Theory of
         Heat Engines, Rigorously Constructed upon the
         Foundation Laid by S. Carnot and F. Reech.**
         New York: Springer, 1977.

         Argues that a classical thermodynamics can be
         built on the 19th-century theories of the heat
         engine put forward by S. Carnot and F. Reech.
         Uses mathematical and technical arguments.

1260.   Walker, G. **Stirling Engines.** New York: Oxford
         University Press, 1980.

         Includes a discussion of the history of the
         Stirling engine and the role of thermodynamics
         in its development.

1261.   Wilson, S.S. "Sadi Carnot." **Scientific
         American** 245 (February 1981): 134-44.

         Argues that Carnot's cycle was not real.
         Concludes that the intention of his work was not
         theoretical but a promotion of the wider appli-
         cation of heat engines.

See also: 43, 98, 123, 137, 193, 229, 272, 290, 308,
314, 315, 316, 321, 336, 337, 357, 358, 367, 368, 370,
371, 372, 380, 385, 645, 680, 807, 1068, 1483.

CHAPTER VIII: FLUID MECHANICS

HYDRAULICS, HYDRODYNAMICS, WATERWHEELS AND TURBINES

Primary Sources

1262.   Alembert, Jean Lerond d'. **Essai d'une nouvelle théorie de la résistance des fluides.** Paris: David, 1752.

Fundamental work on the resistance of fluids by a leading French theorist.

1263.   _____. **Traité de l'équilibre et du mouvement des fluides, pour servir de suite au traité de dynamique.** Paris: David, 1744.

Applies the theory of mechanics to fluids.

1264.   Aubuisson de Voisin, Jean Francois d'. **Traité d'hydraulique, à l'usage des ingénieurs.** Paris: F.G. Levrault, 1834.

Work on hydraulics by a leading French theoretician. Translated into English by Joseph Bennett as **A Treatise on Hydraulics for the Use of Engineers.** Boston: Little, Brown, 1852. Both English and French versions went through several editions.

1265.   _____. **Traité du mouvement de l'eau dans les tuyaux de conduite, à l'usage des ingénieurs.** 2nd edition. Paris: F.G. Levrault, 1836.

Work on the movement of water in pipes.

1266.   Baker, Thomas. **The Principles and Practice of Statics and Dynamics, Embracing a Clear Development of Hydrostatics, Hydrodynamics, and Pneumatics.** London: John Weale, 1851.

Applies the principles of mechanics to the
problems of fluids. Went through five printings.

1267. Bélidor, Bernard Forest de. **Architecture hy-
       draulique, ou l'Art de conduire, d'élever, et
       de menager les eaux pour les différents be-
       soins de la vie.** 4 volumes. Paris: C.A.
       Jombert, 1737-53.

       Early work on attempts to calculate the effi-
       ciency of waterwheels. Concludes, wrongly, that
       undershot wheels are more efficient than over-
       shot wheels. Translated into German.

1268. Bernoulli, Daniel. **Hydrodynamica.** Argentorati:
       J.R. Dulseckeri, 1738.

       Fundamental theoretical work on hydrodynam-
       ics. Presents a scientific analysis of water-
       wheels. Applies the principle of vis viva to
       hydrodynamics.

1269. _____. **Principes hydrostatiques et mécaniques.**
       Paris: Bachelier, 1810.

       Fundamental work on hydrostatics. Originally
       published by the Académie des Sciences, in 1757.

1270. Bernoulli, Jean (Johann). **Hydraulics.** Trans-
       lated by Thomas Carmody and Helmut Kobut.
       Introduction by Hunter Rouse. New York:
       Dover, 1968.

       Translation from the 1743 Latin edition.
       Also includes a translation of Daniel
       Bernoulli's **Hydrodynamica** of 1738 (item 1268).

1271. Borda, Jean Charles. "Sur les roues hy-
       drauliques." **Mémoires de l'Académie des sci-
       ences.** Paris, 1767 (published 1770).

       Attempts to reconcile the discoveries of de
       Parcieux (item 1305) with theory. Applies the
       concept of vis viva to waterwheels.

1272. Bossut, Charles. **Traité élémentaire
       d'hydrodynamique.** Paris: C.-A. Jombert,
       1771.

One of the earliest texts to introduce fluid mechanics to engineers. Contained new experimental data. Expanded two-volume version published in 1786-87 under the title **Traité théorique et expérimental d'hydrodynamique.**

1273.   Coulomb, Charles Augustin de. **Recherches sur les moyens d'exécuter sous l'eau toutes sortes de travaux hydraulique sans employer aucun épuisement.** Paris: C.-A. Jombert, 1779.

Early work on hydraulic engineering by a leading engineering scientist.

1274.   Desaguliers, John Theophilus. **Being a Treatise on Hydrostaticks.** London: J. Senex, 1718.

Early work on hydrostatics by an 18th-century natural philosopher. Had applications to engineering.

1275.   DuBuat, Pierre Louis Georges. **Principles d'hydraulique; ouvrage dans lequel on traité du mouvement de l'eau.** Paris: L'Imprimérie de Monsieur, 1779.

Important 18th-century work on hydraulics. Includes experiments conducted by the author. Supplements the work done by Bossut (item 1272). Later editions published in 1786 and 1816.

1276.   Emerson, James. **Treatise on the Manner of Testing Water-Wheels and Machinery.** Lowell, Mass.: Stone & Huse, 1872.

Provides an example of the systematic testing of waterwheels. Went through several editions. Includes work on water turbines.

1277.   Euler, Johann Albrecht. **Enodatio qvaestionis: Qvomodo vis aqvae alivsve flvidi cvm maximo lvero ad molas civcvm agendas aliave opera perficienda impendi possit?** Göttingen: Svmtibvs Dan. Frid. Kvbleri, 1754.

Theoretically establishes, through the use of rational mechanics and the use of <u>vis viva</u>, that the overshot wheel is more efficient than the undershot.

1278.   Ewbank, Thomas. **A Descriptive and Historical**
        Account of Hydraulic and Other Machines for
        Raising Water, Ancient and Modern: with Ob-
        servations on the Progressive Development of
        the Steam Engine. 2nd edition. New York:
        Greeley & McElrath, 1847.

        Provides a detailed and inclusive account of
        hydraulic devices. Makes important links be-
        tween hydraulic machines and the development of
        the steam engine. Reprinted in a modern edition
        in New York by Arno Press in 1972.

1279.   Eytelwein, Johann Albert. **Handbuch der Hydro-**
        statik. Berlin: G. Reimer, 1826.

        Important study of hydrostatics by the direc-
        tor of the Berlin Bauakademie.

1280.   Fanning, John Thomas. **A Practical Treatise on**
        Hydraulic and Water-Supply Engineering: Re-
        lating to the Hydrology, Hydrodynamics, and
        Practical Construction of Water-Works, in
        North America. New York: D. Van Nostrand,
        1877.

        Useful work which went through eighteen edi-
        tions.

1281.   Flamant, Alfred Aimé. **Mécanique appliquée hy-**
        draulique. Paris: Baudry, 1891.

        One of the first textbooks on the new mathe-
        matical theory of hydrodynamics based on the
        work of Reynolds and Stokes.

1282.   Forchheimer, Phillipp. **Hydraulik.** Leipzig and
        Berlin: B.G. Teubner, 1914.

        Important compilation and commentary on hy-
        draulics.

1283.   Fourneyron, Benoît. "Mémoire sur l'application
        en grand, dans les usines et manufactures, de
        turnines hydrauliques ou roues à palettes
        courbes de Bélidor." **Bulletin des Société**
        d'encouragement pour l'industrie nationale
        33 (1834): 3-17; 49-61; 85-96.

Describes the development of the Fourneyron water turbine. Influenced by the work of Lazare Carnot and Jean Charles Borda (items 1000, 1271).

1284. Francis, James B. **Lowell Hydraulic Experiments.** Boston: Little, Brown, 1855.

Presents the results of experiments conducted at Lowell, Massachusetts on water turbines. Develops a set of design principles for turbines.

1285. Freeman, John Ripley. **Hydraulic Laboratory Practice.** New York: The American Society of Mechanical Engineers, 1929.

Translation and expansion of a study on European hydraulic theory and practice, prepared under the auspices of the Verein Deutscher Ingenieure. Influenced 20th-century engineering education and research.

1286. Genêt, Edmond Charles Edouard. **Memorial on the Upward Forces of Fluids, and Their Applicability to Several Arts, Sciences, and Public Improvements.** Albany: Packard & Van Benthuysen, 1825.

Popular account of the practical application of the science of hydrodynamics.

1287. Gibson, Arnold Hartley. **Hydraulics and Its Applications.** London: A. Constable & Co., 1908.

Popular book on hydrostatics theory, hydraulics, and hydraulic machinery. Went through numerous editions.

1288. Guényveau, André. **Essai sur la science des machines, dans lequel on traité des moteurs, des roues hydrauliques, des machines à colonne d'eau, du belier hydraulique, des machines à vapeur, des hommes et des animaux.** Paris: Bachelier, 1810.

Provides a science of machinery with an emphasis on water powered machines. Includes analysis of human, animal and steam powered machines.

1289.   Helmholtz, Hermann Ludwig Ferdinand von. **Zwei**
        hydrodynamische abhandlungen. Leipzig: W.
        Engelmann, 1896.

        Contains reprints of two important papers on
        hydrodynamics. The first, "Über wirbelbewegun-
        gen," was originally published in 1858 in
        Crelle-Borchardt, Journal für die reine und
        angewandte mathematik, and deals with vortex mo-
        tion in a perfect fluid. The second, "Über dis-
        continuirliche flüssigkeitbewegungen" was origi-
        nally published in 1868 in **Monatsberichte d.**
        Königl. Akademie der Wissenschaften zu Berlin,
        and concerns discontinuous flow. Both papers
        influenced the development of aerodynamics.

1290.   Humphreys, Andrew A., and H.L. Abbot. **Report**
        Upon the Physics and Hydraulics of the
        Mississippi River; Upon the Protection of the
        Alluvial Region Against Overflow and Upon the
        Deepening of the Mouths: Based on Surveys and
        Investigations Submitted to the Bureau of To-
        pographical Engineers. Philadelphia: J.B.
        Lippincott & Co., 1861.

        Valuable contribution to the hydraulic theory
        of rivers. Formed the basis of navigational im-
        provements of the Mississippi. Translated into
        several foreign languages.

1291.   Kelvin, William Thomson, Baron. "Notes on Hy-
        drodynamics." **Cambridge and Dublin Mathemat-**
        ical Journal 2 (1847): 282; 3 (1848): 89; 4
        (1849): 90.

        Three papers on hydrodynamics: the first, "On
        the Equation of Continuity"; the second, "On the
        Equation of the Boundary Surface"; and the
        third, "On the Vis Viva of a Liquid in Motion."
        Written as part of a series co-authored with
        Stokes (item 1323). Helped to present the ideas
        of hydrodynamics to students. Influenced work
        in naval architecture.

1292.   _____. "On Vortex Motion." **Transactions of**
        the Royal Society of Edinburgh 25 (1869):
        217-260.

        Important paper on vortex motion. Influenced
        work in fluid mechanics and aerodynamics.

1293. Lamb, Horace. **Hydrodynamics.** Cambridge:
       Cambridge University Press, 1895.

       One of the first textbooks to make available
       the new theories of hydrodynamics based on the
       work of Reynolds, Stokes, and Rayleigh. Went
       through several editions.

1294. _____. **A Treatise on the Mathematical Theory
       of the Motion of Fluids.** Cambridge:
       Cambridge University Press, 1879.

       Based on lectures given at Trinity College,
       Cambridge in 1874.

1295. Lardner, Dionysius. **A Treatise on Hydrostatics
       and Pneumatics.** London: Longman, Rees, Orme,
       Brown and Green, 1831.

       Popular work on the science of hydrostatics.
       Helped introduce this science to mechanics.

1296. Lea, Frederick Charles. **Hydraulics for Engi-
       neers and Engineering Students.** London: E.
       Arnold, 1909.

       Popular textbook on hydraulics. Went through
       several editions.

1297. McNown, John S., ed. **Classic Papers in Hy-
       draulics.** New York: American Society of
       Civil Engineers, 1982.

       Contains classic papers in hydraulics pre-
       pared by the hydraulic division of the American
       Society for Civil Engineers.

1298. Mariotte, Edme. **The Motion of Water and Other
       Fluids.** Translated by John T. Desaguliers.
       London: J. Senex & W. Taylor, 1718.

       Originally published in Paris in 1686. Early
       text on the effect of the impact of water on
       waterwheel blades. Presents measurements of the
       static weight that counterbalanced flowing
       water. Focused on the undershot wheel. Had
       considerable influence on later works.

1299. Merriman, Mansfield. **A Treatise on Hydraulics.**
       New York: John Wiley & Sons, 1889.

Textbook on hydraulics by a professor of civil engineering at Lehigh University. Made hydraulic theory available to American engineers. Went through several printings.

1300. Morin, Arthur Jules. **Expériences sur les roues hydrauliques à aubes planes et sur les roues hydrauliques à augets.** Metz: Thiel, 1836.

Important work on an experimental study of waterwheels.

1301. _____. **Expérience sur les roues hydraliques à axe vertical appelées turbines.** Paris: L. Mathias, 1838

Experimental study of waterwheels and turbines. American edition translated by Ellwood Morris for the **Journal of the Franklin Institute** in 1843.

1302. Moseley, Henry. **A Treatise on Hydrostatics and Hydrodynamics: For the Use of Students in the University.** Cambridge: T. Stevenson, 1830.

Early textbook on the theory of fluid mechanics.

1303. Navier, Claude Louis Marie Henri. "Mémoire sur l'écoulement des fluides élastiques dans les vases et les tuyaux de conduite." **Mémoires de l'Académie royale des sciences** 2nd series 9 (1830): 311-78.

Analysis of fluid flow in pipes by a leading French theoretician.

1304. _____. "Mémoire sur les lois du mouvement des fluides." **Mémoires de l'Académie royale des sciences** 2nd series 6 (1827): 389-440.

Important work on the laws of hydrodynamics.

1305. Parcieux (Deparcieux), Antoine de. "Mémoire dans lequel on demontre que l'eau d'une chute destinée à faire mouvoir quelque machine, moulin au autre, peut toujours produire beaucoup plus d'effect en agissant par son poids qu'en agissant par son choc, & que les roues a pots qui tournent lentement, produisent plus d'effect que celles qui tournent vite,

relativement aux chutes & aux dépenses."
Mémoires de l'Académie des sciences. Paris,
1754 (published 1756).

One of the first works to recognize that in
waterwheels, gravity was a more efficient power
than impact. Confirmed his analysis with experiments.

1306.   Pardies, Ignace-Gaston. **Statique, ou la science
des forces mouvantes.** Paris, 1673.

Early theoretical study of the mouvement of
bodies in fluids.

1307.   Parent, Antoine. "Sur la plus grande perfection
possible des machine, étant donnée une machine qui ait pour puissance motrice quelque
corps fluide que ce soit, comme, par example,
l'eau, le vent, la flamme, &c." **Mémoires de
l'Académie des sciences.** 2nd edition. Paris,
1704 (published 1772).

One of the first scientific analyses of the
undershot waterwheel. Uses Galilean mechanics
and calculus. Assumed, incorrectly, that results on undershot wheels could be generalized
to all waterwheels. Influential work.

1308.   Poisson, Siméon Denis. "Mémoire sur l'équilibre
des fluides." **Mémoires de l'Académie royal
des sciences** 2nd series 9 (1830): 1-88.

Paper on hydraulics by a leading French theorist and member of the faculty of the École
Polytechnique.

1309.   _____. "Mémoire sur le mouvement de deux fluides élastiques superposes." **Mémoires de
l'Académie royale des sciences** 2nd series
10 (1831): 317-404.

Paper on the hydrodynamics of superimposed
fluids.

1310.   Poncelet, Jean Victor. **Mémoire sur les roues
hydrauliques à aubes courbes, mués par-
dessous, suivi d'expériences sur les effets
mécaniques de ces roues.** Metz: M.V. Thiel,
1827.

Important work on the waterwheel based on the
theories of Lazare Carnot (see item 1000).

1311.   _____. "Mémoire sur les roues verticales à
palettes courbes mues par en dessous, suivi
d'expériences sur les effets mécaniques de
ces roues." **Annales de chimie et de physique**
30, series 2 (1825): 136-88; 225-57.

Presents a new design for a waterwheel, known
as the Poncelet Wheel, based on the theories of
Lazare Carnot and Jean Charles Borda (items
1000, 1271).

1312.   _____. "Théorie des effets mécaniques de la
turbine Fourneyron." **Mémoires de l'Académie
royal des sciences** 2nd series 27 (1840): 3-
36.

Study of the Fourneyron water turbine.

1313.   Prony, Gaspard Clair François Marie Riche, Baron
de. **Nouvelle architecture hydraulique, contenant l'art d'élever l'eau au moyen de dif-
férentes machines, de construire dans ce fluide, de la diriger, et généralement de
l'appliquer, de diverses manières, aux be-
soins de la société.** 2 volumes. Paris:
Firmin Didot, 1790-96.

Detailed description of 18th-century hydraulic machinery. First volume used calculus
while the second volume was more descriptive and
nonanalytical. Based on Lagrange's mechanics
but with a more practical orientation. Also describes steam engines. Prony was director of
the École des Ponts et Chaussées.

1314.   _____. **Recherches physico-mathématiques sur la
théorie des eaux courantes.** Paris: L'Impr.
impériale, 1804.

Mathematical theory of fluids by a leading
French theoretician.

1315.   Rateau, Auguste. **Traité des turbo-machines.**
Paris: V.C. Dunod, 1900.

Treatise on turbines by a noted engineering
scientist.

1316.   _____, and D. Eydoux, and M. Gariel. **Turbines
         hydrauliques.** Paris: J.B. Baillière, 1926.

         Major encyclopaedic work on the theory of hy-
         draulic turbines.

1317.   Redtenbacher, Ferdinand Jacob. **Theorie und Bau
         der Turbinen und Ventilatoren.** Mannheim: F.
         Bassermann, 1844.

         Important work on turbines by the professor
         of mechanical engineering at Karlsruhe Polytech-
         nic.

1318.   _____. **Theorie und Bau der Wasser-Rader.**
         Mannheim: F. Bassermann, 1846.

         Important study of waterwheels.

1319.   Saint-Venant, Adhémar Jean Claude Barré de. **Ré-
         sistance des fluides: Considérations his-
         toriques, physiques et practiques relatives
         au problème de l'action dynamique mutuelle
         d'un fluide et d'un solide, spécialement dans
         l'état de permanence supposé acquis par leurs
         mouvements.** Paris: F. Didot, 1888.

         Important work on the resistance of fluids by
         a leading figure in the theory of elasticity.

1320.   Smeaton, John. **Experimental Enquiry Concerning
         the Natural Powers of Wind and Water to Turn
         Mills and Other Machines Depending on a Cir-
         cular Motion.** London: I. and J. Taylor,
         1794.

         Presents important early experimental studies
         on waterwheels by a famous British engineer.
         What little theory that is used is based on
         Newtonian science. Uses reduced scale test ap-
         paratus. Concludes that the theoretical effi-
         ciency of an overshot waterwheel is 1 while the
         efficiency of an undershot waterwheel is 1/2.
         Early account published in the **Philosophical
         Transactions of the Royal Society of London** 51
         (1759): 100-174. Translated into French.

1321.   Smith, Hamilton, Jr. **Hydraulics. The Flow of
         Water through Orifices, over Weirs, and
         through Open Conduits and Pipes.** New York:
         John Wiley & Sons, 1886.

Provides a comprehensive treatment of the subject.

1322. Stokes, George Gabriel. "Report on Recent Researches on Hydrodynamics." **Report of the British Association for the Advancement of Science** (1846): 1-31.

Influential summary of the theory of hydrodynamics. Includes some discussion of the theory of viscous flow.

1323. _____. "Notes on Hydrodynamics." **Cambridge and Dublin Mathematical Journal** 3 (1848): 121-27; 3 (1848): 209-33; and 4 (1849): 219-40.

Three papers on hydrodynamics: the first, "On the Dynamical Equations"; the second, "Demonstration of a Fundamental Theorem"; and the third, "On Waves." Series of papers co-authored with William Thomson, Lord Kelvin (see item 1291). Helped introduce students to the theory of hydrodynamics. Influenced the development of naval architecture.

1324. Switzer, Stephen. **An Introduction to a General System of Hydrostaticks and Hydraulics, Philosophical and Practical.** London, 1729.

Early attempt to discover whether overshot or undershot waterwheels were more efficient. Focused on the role of impact rather than weight of water.

1325. Tredgold, Thomas, ed. **Tracts on Hydraulics.** London: J. Taylor, 1826.

Contains the results of experimental studies by John Smeaton, Giovanni Venturi, Johann Eytelwein, with a summary by Thomas Young.

1326. Venturi, Giovanni Battista. **Experimental Enquiries Concerning the Principle of Lateral Communication of Motion in Fluids: Applied to the Explanation of Various Hydraulic Phenomena.** London: J. Taylor, 1799.

Fundamental work on the dynamics of fluid flow. Originally published in French in Paris in 1797.

1327.   Weisbach, Julius. **Die Experimental-Hydraulik. Eine Anleitung zur Ausführung hydraulisher Versuche im Kleinen, nebst Beschreibung der hierzu nöthigen Apparate und Entwickelung der wichtegsten Grundformeln der Hydraulik, so wie Vergleichung der durch diese Apparate gefundenen Versuchsresultate mit der Theorie und mit den Erfahrungen im Grossen.** Freiberg: J.G. Engelhardt, 1855.

Important work on hydraulics by a professor at the Bergakademie in Freiberg.

1328.   Wood, De Volson. **Theory of Turbines.** New York: J. Wiley & Sons, 1895.

Important textbook on turbines. Second edition published in 1896 as **Turbines, Theoretical and Practical.**

1329.   Zeuner, Gustav Anton. **Vorlesungen über Theorie der Turbinen. Mit vorbereitenden Üntersuchungen ausder technischen Hydraulik.** Leipzig: A. Felix, 1899.

Work on the theory of the turbine. Translated into French in 1905.

See also: 388, 389, 391, 393, 399, 406, 411, 615, 689, 700, 701, 702, 714, 716, 727, 744, 759, 774, 779, 790, 799, 1007, 1011, 1013, 1036, 1377, 1381, 1385, 1400, 1404, 1412, 1419, 1461.

Secondary Sources

1330.   Apmann, Robert P. "A Case History in Theory and Equipment: Fluid Flow in Bends." **Isis** 55 (1964): 427-34.

Discusses the work on fluid mechanics of J.J. Thomson and Joseph Boussinesq.

1331. Bergia, S., and P. Fantanzzini. "La descrizione
       dei fenomeni meccanici in termini energetici
       nell'opera di John Smeaton. Parte 1: Dalle
       route ad acqua ai teoremi dell'impulso e
       delle forze vivre. Parte 2: Un modello per
       l'utro anelastico ed il suo sontrollo
       sperimentale." **Giornale de Fisica** 22
       (1981): 295-310; and 23 (1983): 59-73.

       Analyzes the work of John Smeaton in the de-
       velopment of the theory of the waterwheel.

1332. Bied-Charreton, René. "L'Utilization de
       l'énergie hydraulique: Ses origines, ses
       grandes étapes." **Revue d'Histoire des Sci-
       ences** 8 (1955): 53-72.

       Surveys the use of hydraulics as a source of
       energy during the 18th, 19th and 20th centuries.
       Emphasizes French contributions.

1333. Binnie, G.M. **Early Victorian Water Engineers.**
       London, 1981.

       Includes a discussion of the role of theory
       in the development of water supply systems.

1334. Biswas, Asit K. **History of Hydrology.** New
       York: American Elsevier, 1970.

       Presents a survey of hydrology from 600 B.C.
       to the 19th century. More factual than analyti-
       cal. Discusses the development of quantitative
       hydrology beginning in the 17th century.

1335. _____. "A History of Hydrology in the 19th
       Century." **Water Power** 21 (1969): 16-21.

       Surveys the major contributions to hydrology
       during the 19th century.

1336. Braun, Hans-Joachim. "Technische Neuerungen um
       die Mitte des 19. Jahrhunderts: Das Beispiel
       der Wasserturnbinen." **Technikgeschichte** 46
       (1979): 285-305.

       Discusses the role of engineering science in
       the diffusion of turbine technology during the
       19th century.

1337.   Constant, Edward W., II. "Scientific Theory and
         Technological Testability: Science, Dynamo-
         meters, and Water Turbines in the 19th Cen-
         tury." **Technology and Culture** 24 (1983):
         183-98.

         Argues that technological testing was a key
         element in the development and diffusion of
         19th-century water turbines, and that the test-
         ing hardware and procedures embodied scientific
         information and provided a major mechanism for
         the interaction of science and technology. Be-
         lieves that testing procedures can became inde-
         pendent of specific systems and become new tech-
         nological traditions.

1338.   Cotton, Gordon. **A History of the Waterways Ex-
         periment Station, 1929-1979.** Vicksburg,
         Miss.: Corps of Engineers, 1979.

         Discusses the Army Corps of Engineers' role
         in studying hydrology through the construction
         of models. Discusses experiments on flood con-
         trol, soil testing, levee and dam construction.
         Describes experiments conducted for the military
         during World War II and for NASA's Apollo pro-
         ject. No discussion of environmental, social or
         economic issues.

1339.   Crozet-Fourneyron, Marcel. **Invention de la tur-
         bine: Historique suivi d'une note sur un rég-
         ulateur à mouvement louvoyant applicable aux
         turbines hydrauliques.** Paris: Beranger,
         1927.

         Discusses the invention of the turbine and
         the control of motion in the turbine.

1340.   Danilevskii, Viktor V. **History of Hydro-Engi-
         neering in Russia before the 19th Century.**
         Jerusalem: Israel Program for Scientific
         Translation, 1968.

1341.   Durand, W.F. "The Development of Our Knowledge
         of the Laws of Fluid Mechanics." **Science** 78
         (1933): 342-51.

         Survey of the development of fluid mechanics
         by a leading contributor to the field.

1342.   _____. "The Pelton Water Wheel." **Mechanical
        Engineering** 61 (1939): 447-54; 511-18.

        Discusses the development of the Pelton
        waterwheel.

1343.   Frazier, Arthur H. **Water Current Meters in the
        Smithsonian Collections of the National
        Museum of History and Technology.**
        Washington, D.C.: Smithsonian Institution
        Press, 1974.

        Describes the development of stream flow mea-
        surement from 1500 B.C. to the present. Dis-
        cusses the application of the meters to problems
        arising in hydraulics.

1344.   Gandillot, Maurice. **Note sur une illusion de
        relativité.** Paris: Gauthier-Villars, 1913.

        Discusses the resistance of fluids.

1345.   Gibson, A.H. **Osborne Reynolds and His Work in
        Hydraulics and Hydrodynamics.** London: The
        British Council, 1948.

        Discusses the contributions of a major figure
        in fluid dynamics and the person after whom the
        Reynolds Number is named.

1346.   Gordon, Robert B. "Hydrological Science and the
        Development of Waterpower for Manufacturing."
        Technology and Culture 26 (1985): 204-235.

        Investigates the problems faced by New
        England manufacturers and the solutions found in
        developing waterpower. Argues that the scien-
        tific concepts of hydrology and geology were
        formulated too late to help design waterpower
        systems during their most rapid growth. As a
        result, steam power, which was better under-
        stood, was used for manufacturing. Argues that
        both published works and the scientific under-
        standing of hydrology needed to sustain indus-
        trial growth were lacking.

1347.   Guillerme, André. **Les temps de l'eau: La cité,
        l'eau et les techniques, nord de la France,
        fin IIIe--début XIXe siècle.** Seyssel: Champ
        Vallon, 1983.

Presents a study of the role of water in ur-
ban centers in northern France from late Roman
times to the 19th century. Provides some discussion of contemporary scientific theories such
as the hydrologic cycle.

1348.   Hunter, Louis C. **A History of Industrial Power
in the United States.** Volume 1: **Waterpower
in the Century of the Steam Engine.**
Charlottesville: University Press of Virginia
for the Eleutherian Mills-Hagley Foundation,
1979.

Discusses the rise and decline of water power
in the United States from the use of mill power
to its comptetion with steam power. Emphasis
placed on the relationship between technology
and economic history. Includes a detailed description of the development of the water turbine. Shows that turbine technology made great
improvements in efficiency and design even in
the face of economic decline.

1349.   _____. "Les Origines des turbines Francis et
Pelton: Développement de la turbine hydraulique aux États-Unis de 1820 à 1900."
Revue d'Histoire des Sciences 17 (1964):
209-242.

Discusses the development of the hydraulic
turbine in America. Compares the work of
Francis and Pelton.

1350.   Layton, Edwin T., Jr. "Benjamin Tyler and the
American Antecedents of the Hydraulic Turbine." **Tools & Technology** 5 (1983): 15-18.

Traces the development of the water turbine
back to the 1780s with the development of
Benjamin Tyler's wry-fly waterwheel. Argues
that it has all of the main features of the
modern turbine.

1351.   _____. "The Industrial Evolution of America:
Energy in the Age of Water and Wood."
"Energy in American History." Edited by
Arthur Donovan. **Materials and Society** 7
nos. 3/4 (1983): 249-63.

Discusses the shift from low-power water mills to the use of the water turbine during the late 18th and early 19th centuries. Shows that early improvements of waterwheels and reaction wheels were conscious attempts to create a scientific technology. Argues that the vernacular, empirical science of American millwrights could be considered just as scientific as the mathematical theory of France, but simply reflected different scientific traditions.

1352.   _____. "Scientific Technology, 1845-1900: The Hydraulic Turbine and the Origins of American Industrial Research." **Technology and Culture** 20 (1979): 64-89.

Traces four stages of turbine development: (1) the application of highly simplified science to the reaction wheel, (2) the creation of a body of engineering science based on experimental research applied to the Fourneyron turbine, (3) the expansion of experimental industrial research associated with large-scale production of the Francis turbine, (4) the establishment of industrial research laboratories creating a theoretical base for turbine science associated with the needs of hydroelectric power. Argues that engineering science is based on design principles and teleology rather than generality and exactitude as is pure science. Important case study.

1353.   Nemenyi, P.F. "The Main Concepts and Ideas of Fluid Dynamics in Their Historical Development." **Archive for History of Exact Sciences** 2 (1962): 52-86.

Detailed internalistic survey of the history of fluid dynamics.

1354.   Noir, Dominique. "La première loi de similitude de la mécanique des fluides." **Revue de l'Histoire des Sciences et de leurs Applications** 25 (1972): 271-74.

Traces similarity laws used by engineers back to Aristotle and compares them to a law on the drag formula of a sphere given by George G. Stokes.

1355. Redondi, Pietro. "Along the Water: The Genius
       and the Theory: D'Alembert, Condrocet and
       Bossut and the Picardy Canal Controversy."
       History and Technology 2 (1985): 77-110.

       Discusses the role of d'Alembert, Condrocet
       and Bossut in the controversy over creating an
       underground canal from the North Sea to central
       France.

1356. Reuss, Martin. "Andrew A. Humphreys and the De-
       velopment of Hydraulic Engineering: Politics
       and Technology in the Army Corps of Engi-
       neers, 1850-1950." **Technology and Culture**
       26 (1985): 1-33.

       Argues that political considerations shaped
       the application of science to technology in the
       hydraulic projects of the Army Corps of Engi-
       neers.

1357. Reynolds, Terry S. "Scientific Influences on
       Technology: The Case of the Overshot Water-
       wheel, 1752-1754." **Technology and Culture**
       20 (1979): 270-95.

       Describes how John Smeaton, Antoine de
       Parcieux and Johann Albrecht Euler established
       the superior efficiency of the overshot-gravity
       waterwheel as compared to the undershot-impulse
       wheel. Argues that the influence of science on
       technology was largely indirect. Shows that the
       traditional work of practical engineers (Smeaton
       and de Parcieux) using intuition, analogy, and
       experimentation, had more influence on water-
       wheel technology than did the work of scientists
       (Euler) using the direct application of mathe-
       matics to technology.

1358. _____. **Stronger than A Hundred Men: A History
       of the Vertical Water Wheel.** Baltimore:
       Johns Hopkins University Press, 1983.

       Describes how the theoretical and empirical
       investigations of waterwheels, since the 16th
       century, led to a new scientific theory of de-
       sign for waterwheels in the 18th and 19th cen-
       turies. Shows that a lack of scientific under-
       standing led to deficiencies in pre-19th-century
       waterwheels. Argues that science and the devel-
       opment of water power converged at critical re-

search sites during the 18th century. Analyzes
the relationship between economics and scientific design. Important study.

1359. Rouse, Hunter. **Hydraulics in the United States,
       1776-1976.** Iowa City: Institution of Hy-
       draulic Research, University of Iowa, 1976.

       Focuses on the scientific development of hy-
       draulics in the United States. Presents brief
       biographical sketches of the major researchers
       in hydraulics. Emphasizes the twentieth cen-
       tury. Provides brief descriptions of some major
       waterworks in Boston, New York and Philadelphia.
       Provides a valuable tool for further research.

1360. _____, and Simon Ince. **History of Hydraulics.**
       Ames, Iowa: Iowa Institute of Hydraulic Re-
       search, State University of Iowa, 1957.

       Surveys the history of hydraulics with empha-
       sis on European developments since 1600.

1361. Shaw, John. **Water Power in Scotland, 1550-1870.**
       Edinburgh: John Donald Publishers, 1984.

       Extensive survey of water power in Scotland.
       Emphasis is on narrative description rather than
       the application of theory to hydraulics.

1362. Smith, Norman A. F. **Man and Water: A History of
       Hydro-Technology.** New York: Charles
       Scribner's Sons, 1976.

       Surveys the entire history of hydraulic tech-
       nology from ancient civilizations to modern
       times. Presents a discussion of the evolution
       of waterpower technology in part three and a
       broad discussion of hydrotechnology in part
       four.

1363. _____. "The Origins of the Water Turbine."
       Scientific American 242 (January 1980): 138-
       48.

       Discusses the improvements in water wheels
       that led to the development of the water tur-
       bine.

1364. _____. "The Origins of the Water Turbine and
       the Invention of Its Name." **History of Tech-
       nology** 2 (1977): 215-59.

       Studies the history of the water turbine be-
       fore Fourneyron. Argues that the claim that
       Fourneyron invented the water turbine is mis-
       leading.

1365. Szabo, Istvan. "Über die sog 'Bernoulische
       Gleichung' der Hydromechanik: Die Stromfaden-
       theorie Daniel und Johann Bernoullis." **Tech-
       nikgeschichte** 37 (1970): 27-64.

       Study of Daniel and Johann Bernoulli's con-
       tribution to the theory of hydromechanics.

1366. Tokaty, G.A. **A History and Philosophy of Fluid-
       mechanics.** Henley-on-Thames: G.T. Foulis,
       1971.

       The first half of the book provides a history
       of the standard developments in fluid mechanics.
       The second half of the book focuses on Russian
       and Soviet contributions. Many of the Russian
       sources are not easily available in other
       English sources.

1367. Trevena, David H. "Reynolds on the Internal Co-
       hesion of Liquids." **American Journal of
       Physics** 47 (1979): 341-345.

       Discusses the contribution of Osborne
       Reynolds to fluid mechanics.

1368. Truesdell, Clifford Ambrose. "Experience, The-
       ory, and Experiment." **Bulletin of the State
       University of Iowa Studies in Engineering**
       no. 36 (1956): 3-18.

       Presents a methodological study of fluid me-
       chanics based on historical examples.

1369. _____. "Notes on the History of the General
       Equations of Hydrodynamics." **American Mathe-
       matical Monthly** 60 (1953): 445-58.

       Describes an exhibition of books commemorat-
       ing the bicentennial of the basic equations of
       hydrodynamics.

1370. _____. "Rational Fluid Mechanics, 1687-1765."
Editor's introduction to **L. Euleri Opera
Omnia** 2nd series 12 (1954): ix-cxxv.

Useful survey of the early developments in
fluid dynamics. Emphasizes the role of Euler.

1371. _____. "I. The First Three Sections of Euler's
Treatise on Fluid Mechanics (1766); II. The
Theory of Aerial Sound, 1687-1788; III. Rational Fluid Mechanics, 1765-1788." Editor's
introduction to **L. Euleri Opera Omnia** 2nd
series 13 (1956): vii-cxviii.

Useful articles on 18th-century fluid mechanics.

1372. Wilson, Paul N. "The Waterwheels of John
Smeaton." **Transactions of the Newcomen Society** 30 (1955-57): 25-48.

Describes Smeaton's work on the theory and
practice of waterwheels.

See also: 76, 89, 105, 110, 193, 201, 262, 289, 334,
338, 347, 361, 383, 590, 633, 654, 663, 668, 807, 831,
833, 890, 901, 1052.

NAVAL ARCHITECTURE AND THE THEORY OF WAVES

Primary Sources

1373. Aubin, Nicolas. **Dictionnaire de la marine contenant les termes de la navigation et de
l'architecture navale.** Amsterdam, 1702.

Attempts to rationalize the practice of shipbuilding.

1374. Beaufoy, Mark. **Nautical and Hydraulic Experiments, with Numerous Scientific Miscellanies.**
London: M. Beaufoy, 1834.

Important set of experiments which examined
the role of friction on the movement of ships
through the water.

1375.   Bernoulli, Daniel, and Leonhard Euler, and
        Jacques Mathon de la Cour. **Recherches sur la
        manière de suppléer à l'action du vent sur
        les grands vaisseaux.** Paris: Bachelier,
        1810.

        Application of the principle of force to the
        theory of ship propulsion.

1376.   Bernoulli, Jean. **Essai d'une nouvelle théorie
        de la manoeuvre des vaisseaux avec quelques
        lettres sur le même sujet.** Basel: J.G.
        Konig, 1714.

        Application of scientific principles to de-
        signing ships.   Discusses error in the book of
        B. Renau d'Elissagaray (item 1384).

1377.   Bossut, Charles. **Nouvelles expériences sur la
        résistance des fluides.** Paris: C.-A.
        Jombert, 1777.

        Study of the motion of bodies in fluids done
        for the Académie des Sciences by Bossut,
        d'Alembert and Condrocet.   Bossut was professor
        of mathematics at the Écoles Royale Militaires
        in Mézières. Influenced naval architecture.

1378.   Bouger, Pierre. **De la manoeuvre des vaisseaux
        ou traité de mécanique et dynamique; dans
        lequel on reduit à des solutions très-simples
        les problèmes de marine les plus difficiles,
        qui ont pour objet le nouvement du navire.**
        Paris: H.L. Guerin & L.F. Delatour, 1757.

        Applies mechanics and dynamics to the design
        of ships.   One of the earliest works to apply
        higher mathematics to the problem of the stabil-
        ity of ships.

1379.   _____. **Traité du navire, de sa construction et
        des ses mouvements.** Paris, 1746.

        Presents a method for calculating a ship's
        displacement and its metacenter.

1380.   Bourne, John. **A Treatise on the Screw Pro-
        peller, with Various Suggestions of Improve-
        ment.** London: Longman, Brown, Green, and
        Longman, 1852.

Theoretical and practical treatise on the screw propeller.

1381. Cauchy, Augustin Louis. **Théorie de la propaga-
        tion des ondes à la surface d'un fluide
        pesant d'une profondeur indéfinie.** Paris,
        1827.

Important prize-winning study of the theory of wave motion by a leading French theorist. Influenced naval architecture.

1382. Duhamel du Monceau, Henri-Louis. **Éléments
        d'architecture navale, ou traité pratique de
        la construction des vaisseaux.** Paris, 1754.

One of the first training manuals for naval architects. Attempts to rationalize building practice by the inspector general of the navy.

1383. Durand, William Frederick. **Researches on the
        Performance of the Screw Propeller.**
        Washington, D.C.: Carnegie Institution, 1907.

Report of highly technical testing of screw propellers conducted between 1897 and 1907.

1384. Elissagaray, Bernard Renau d'. **De la Théorie
        de la manoeuvre des vaisseaux.** Paris, 1689.

Early work on the mechanical principles of shipbuilding.

1385. Euler, Leonhard. **Scientia navalis.** 2 volumes.
        Petropoli: Academiae scientiarvm, 1749.

Fundamental work on the science of fluids and naval architecture by a leading theorist.

1386. _____. **Théorie complète de la construction et
        de la maneuvre des vaisseaux.** St. Peters-
        burg: Académie impériale des sciences, 1773.

Practical work on naval architecture by one of the leading 18th-century scientists. Trans-lated into English as **A Compleat Theory of the Construction and Properties of Vessels with Practical Conclusions for the Management of Ships Made Easy to Navigators.** London: P. Elmsley, 1776.

1387.   Fincham, John. **A History of Naval Architecture
        to which Is Prefixed an Introductory Disser-
        tation on the Application of Mathematical
        Science to the Art of Naval Construction.**
        London: Whittaker & Co., 1851.

        Argues for the improvement of naval architec-
        ture through scientific research into shipbuild-
        ing construction.

1388.   Froude, William. "Experiments on the Surface
        Friction Experienced by a Plane Moving
        Through Water." **British Association for the
        Advancement of Science Report** 42 (1872): 118-
        24.

        Important experiments on the theory of waves
        by a noted naval architect.

1389.   _____. "On Experiments with H.M.S. Greyhound."
        **Transactions of the Institution of Naval Ar-
        chitects** 16 (1874): 36-73.

        Important experiments by a noted naval archi-
        tect.

1390.   _____. "On Some Difficulties in the Received
        View of Fluid Friction." **British Association
        for the Advancement of Science** 29 (1869):
        211-14.

        Study on the theory of fluid friction and its
        role in naval architecture.

1391.   _____. "On the Influence of Resistance upon
        the Rolling of Ships." **Naval Science** 1
        (1872): 411-29.

        Study on the stability of ships.

1392.   _____. "On the Practical Limits of the Rolling
        of a Ship in a Sea-Way." **Transactions of the
        Institution of Naval Architects** 6 (1865):
        175-86.

        Study on the stability of ships.

1393.   _____. **The Resistance of Ships.** Washington,
        D.C.: Government Printing Office, 1888.

Important work on naval architecture. Includes a discussion of the wave-making resistance of ships.

1394. Gray, John MacFarlane. "On Polar Diagrams of
       Stability." **Transactions of the Institution
       of Naval Architects** 16 (1875): 85-88.

Study on the stability of ships by a leading
engineering scientist.

1395. Hoste, Paul. **Théorie de la construction des
       vaisseaux.** Lyon, 1697.

Early study of the relationship between speed
and resistance in fluids. Also provides a study
of the form of ships.

1396. Lardner, Dionysius. **Rudimentary Treatise on Ma-
       rine Engines and Steam Vessels; Together with
       Practical Remarks on the Screw and Propelling
       Power as Used in the Royal and Merchant Navy.**
       2nd edition. London: John Weale, 1852.

Popular work on marine engines and ship
propulsion by an important popularizer of scientific ideas.

1397. **A Naval Encyclopaedia: Comprising a Dictionary
       of Nautical Words and Phrases; Biographical
       Notices, and Records of Naval Officers; Spe-
       cial Articles on Naval Art and Science, Writ-
       ten Expressly for this Work by Officers and
       Others of Recognized Authority in the
       Branches Treated by Them, Together with De-
       scriptions of the Principle Naval Stations
       and Seaports of the World.** Detroit: Gale
       Research, 1971.

Reprint of the 1884 edition published in
Philadelphia.

1398. Peake, James. **Rudiments of Naval Architecture;
       or, an Exposition of the Elementary Princi-
       ples of the Science and the Practical Appli-
       cation to Construction.** London: John Weale,
       1849.

Early work on the application of applied mechanics and hydrodynamics to naval architecture
by an assistant master shipwright at H.M. Dock-
yard, Woolwich.

1399.   Pitot, Henri. **La Théorie de la manoeuvre des
        vaisseaux réduit en pratique ou les principes
        et les règles pour naviguer le plus avan-
        tageusement qu'il est possible.** Paris, 1731.

Based on a new theory of ship handling. Influenced by the work of Jean Bernoulli (item
1376). Translated into English by Edmund Stone
as **The Theory of the Working of Ships, Applied
to Practice.** London: C. Davis, 1743.

1400.   Poisson, Siméon Denis. Mémoire sur la théorie
        des ondes." **Mémoires de l'Académie royale
        des sciences** 2nd series 1 (1818): 71-186.

Important paper on the theory of waves. Influenced naval architecture.

1401.   Rankine, William John Macquorn. "On Plane
        Water-Lines in Two Dimensions." **Philosophi-
        cal Transactions of the Royal Society of
        London** 154 (1864): 369-91.

Investigates the curves suitable for the
water-lines of ships. Introduces the mathematical theory of determining the shape of ships.
Formed the basis of his theory of stream lines.
Influenced naval architecture.

1402.   _____. "On the Action of Waves upon a Ship's
        Keel." **Transactions of the Institution of
        Naval Architects** 5 (1864): 20-34.

Significant contribution to the theory of the
rolling of ships. Extended by papers "On the
Isochronous-Rolling Ships." **Trans. Inst. Naval
Arch.** 5 (1864): 35-37, and "On the Uneasy
Rolling of Ships." **Trans. Inst. Naval Arch.** 5
(1864): 38-42. All were influential on naval
architecture.

1403.   _____. "On the Mechanical Principles of the
        Action of Propellers." **Transactions of the
        Institution of Naval Architects** 6 (1865):
        13-39.

Important paper on the theory of propellers. The first to appear in the **Transactions.** Defined real and apparent slip. Calculated the work done by a propeller and its efficiency.

1404.   _____. "On Stream-Line Surfaces." **Transactions of the Institution of Naval Architects.** 11 (1870): 175-81.

Important paper on stream-line surfaces. Proved fallacious the older methods of finding a solid of least resistance. First paper on stream-line surfaces presented to the Institution of Naval Architects. Reprinted in **Engineering** 9 (1870): 267-68.

1405.   _____. "Stream Lines and Waves in Connexion with Naval Architecture." **Engineering** 10 (1870): 233, 252.

Useful survey of the development of the theory of stream lines and their role in naval architecture. Includes a discussion of the earlier work of John Scott Russell.

1406.   _____, I. Watts, F.K. Barnes, and J.R. Napier. **Shipbuilding, Theoretical and Practical.** London: W. Mackenzie, 1866.

One of the first theoretical works on naval architecture. Contains sections on wave theory, applied mechanics, and the use of steam power.

1407.   Reed, Edward Jones. "On the Stability of Monitors under Canvas." **Transactions of the Institution of Naval Architects** 9 (1868): 198-207.

A leading naval architect predicts the instability of turret ships with low freeboards and masts.

1408.   _____. "The Distribution of Weight and Buoyancy in Ships." **Naval Science** 1 (1872): 236-48.

Study of buoyancy in ships.

1409.   _____. "The Rolling of Ships." **Naval Science**
         1 (1872): 172-216.

         Study of waves' effects on ships.

1410.   _____. "The Stability of the 'Captain,'
         'Monarchy' and Some Other Iron-Clads." **Naval
         Science** 1 (1872): 26-42.

         Study of the stability of iron-clad ships.

1411.   Rees, Abraham. **Rees's Naval Architecture, 1819-
         1820, and Extract from the Cyclopaedia; or
         Universal Dictionary of Arts, Sciences and
         Literature.** Newton Abbot, Devon: David &
         Charles, 1970.

         Reprints of articles from **Rees's Cyclopaedia**
         (1819-20). Provides details on the design and
         building of wooden ships.

1412.   Russell, John Scott. "Experimental Researches
         into the Laws of Certain Hydrodynamical Phe-
         nomena that Accompany the Motion of Floating
         Bodies, and Have Not Previously been Reduced
         into Conformity with the Knowledge of the Re-
         sistance of Fluids." **Transactions of the
         Royal Society of Edinburgh** 14 (1840): 47-
         109.

         Important work on the theory of waves. In-
         fluenced naval architecture.

1413.   _____. **The Modern System of Naval Architec-
         ture.** London: Day & Son, 1864-65.

         Fundamental work on naval architecture. In-
         cludes sections on shipbuilding, marine engi-
         neering and steam navigation.

1414.   _____. "On the Education of Naval Architects
         in England and France." **Transactions of the
         Institution of Naval Architects** 4 (1863):
         163-76.

         Description of the education of naval archi-
         tects in England and France by one of the lead-
         ing British naval architects.

1415.   _____. "On the Solid of Least Resistance."
         **British Association for the Advancement of
         Science Report** 1835, part 2, pp. 107-108.

         Shows that solids shaped like a wave offer
         the least resistance when moving through the
         water.

1416.   _____. "Report of a Committee on the Form of
         Ships." **British Association for the Advance-
         ment of Science Report** 1841, pp. 325-26;
         1842, pp. 104-105.

         Argues that ships should have a wave-line
         shape.

1417.   _____. **Very Large Ships, Their Advantages and
         Defects.** London, 1863.

         Includes an analysis of the problems of sta-
         bility and propulsion of large ships.

1418.   Steel, David. **The Elements and Practice of
         Naval Architecture.** London: C.Whittingham,
         1805.

         Early work on naval architecture. Reprinted
         in a modern edition in London by Sim Comfort As-
         sociation in 1977.

1419.   Stokes, George Gabriel. "On the Steady Motion
         of Incompressible Fluids." **Transactions of
         the Cambridge Philosophical Society** 7
         (1842): 439-54; 465.

         Important paper on the motion of fluids. In-
         troduces the concept of lines of motion, which
         later became known as stream lines. Had signif-
         icant applications in the area of naval archi-
         tecture.

1420.   Thearle, Samuel J.P. **Theoretical Naval Archi-
         tecture: A Treatise on the Calculations In-
         volved in Naval Design.** 2 volumes. New
         York: G.P. Putnam's Sons, [c.1877].

         American edition of a European work on the
         theory of naval architecture by a fellow of the
         Royal School of Naval Architects, London. Used
         at the University of Pennsylvania.

See also: 390, 401, 408, 848, 1291, 1323.

Secondary Sources

1421.   Acerra, Martine. "Les constructeurs de la ma-
        rine (XVIIe-XVIIIe siècle)." **Revue
        d'Histoire des Sciences et de leurs Applica-
        tions** 273 (1985): 283-304.

        Surveys naval architecture during the 17th
        and 18th centuries.

1422.   Anderson, R.C. "Eighteenth Century Books on
        Shipbuilding, Rigging, and Seamanship."
        Mariner's Mirror 33 (1947): 218-225.

        Critical study of 18th-century works.

1423.   Barnaby, K.C. **100 Years of Specialized Ship-
        building and Engineering.** London:
        Hutchinson, 1964.

        Includes a discussion of naval architecture.

1424.   Barnaby, Nathaniel. **Naval Development in the
        Century.** London: W. & R. Chambers, 1902.

1425.   Denny, Archibald. "Fifty Years' Evolution in
        Naval Architecture and Marine Engineering."
        Nature 116 (1925): 468-71.

        Surveys the development of naval architecture
        since 1875.

1426.   Flamm, O. "The Scientific Study of Naval Archi-
        tecture in Germany." **Transactions of the In-
        stitution of Naval Architects** 53, part 2
        (1911).

1427.   Gardiner, Robert. "The Frigate Designs of 1755-
        57." **Mariner's Mirror** 63 (February 1977):
        51-69.

        Discusses the designs of British naval ves-
        sels in the mid-18th century.

1428.   Gilfillan, S.C. **Inventing the Ship.** Chicago:
        Follett, 1935.

Includes a discussion of ship design and the
problems of proplusion. Describes the role of
theory.

1429. Gille, Paul. "La construction navale et les
méthodes scientifiques." **Actes VIIe Congrès
International d'Histoire des Sciences.**
Paris: Hermann, 1953, pp. 313-15.

Discusses the role of the scientific method
in the development of naval construction.

1430. _____. "Les mathématiques et la construction
navale." **Actes VIIIe Congrès International
d'Histoire des Sciences.** Paris: Hermann,
1958, pp. 57-63.

Discusses the role of mathematics in the development of naval architecture.

1431. Hall, A. Rupert. "Architectura Navalis."
Transactions of the Newcomen Society 51
(1979-80): 157-74.

Discusses the history of naval architecture.

1432. House, Derek, ed. **Five Hundred Years of Nautical Science, 1400-1900.** Greenwich: National
Maritime Museum, 1981.

Contains several useful articles on naval architecture.

1433. Maloney, Linda McKee. "A Naval Experiment."
American Neptune 34 (July 1974): 188-96.

Describes an 1832 experiment to construct a
naval schooner without an interior frame.

1434. Middleton, William. "Giovanni Alfonson Borelli
on the Construction of Galleys." **Mariner's
Mirror** 59 (1973): 3-8.

Discusses a paper prepared in 1674 on naval
architecture by the famous physiologist.

1435. Milligan, John D. "Charles Ellet, Naval Architect: A Study in Nineteenth Century Professionalism." **American Neptune** 31 (January
1971): 152-72.

Describes the contributions of Ellet to naval architecture.

1436.   Pollard, Sidney, and Paul Robertson. **The British Shipbuilding Industry, 1870-1914.** Cambridge, Mass.: Harvard University Press, 1979.

Includes a chapter on scientific shipbuilding and the evolution of technical education.

1437.   Prichard, James.   "From Shipwright to Naval Constructor: The Professionalization of 18th-Century French Naval Shipbuilders." **Technology and Culture** 28 (1987): 1-25.

Argues that in shipbuilding the application of science to technology had an effect only after the appearance of institutional and social developments. Maintains that science affected shipbuilding indirectly through the establishment of schools in which science was taught to naval constructors. Concludes that professionalization resulted in shipbuilders' becoming engineers rather than scientists' becoming shipbuilders.

1438.   Sandler, Stanley.   "The Emergence of the Modern Capital Ship." **Technology and Culture** 11 (1970): 576-95.

Discusses the introduction of mastless, seagoing, turreted, ironclad ships into the Royal Navy. Analyzes the debates between Captain Cowper Coles and naval architect E.J. Reed over the stability of turrets on ships.

1439.   Sattiram, Kappal.   "A Tamil Treatise on Shipbuilding during the 17th Century A.D." **Indian Journal of History of Science.** 7 (1972): 16-26.

1440.   Scott, John C.   "The Historical Development of Theories of Wave-Calming Using Oil." **History of Technology** 3 (1978): 163-86.

Describes the theories from antiquity to the present concerning the mechanism involved in wave-calming by the use of oil. Discusses the theory of waves and the theory of surface ten-

sion. Includes a description of the contributions of Osborne Reynolds.

1441.   Smith, Edgar C. "The Centenary of Naval Engi-
        neering." **Transactions of the Newcomen Soci-
        ety** 2 (1921-22): 88-114.

1442.   _____. **A Short History of Naval and Marine En-
        gineering.** Cambridge: Cambridge University
        Press, 1938.

        Useful survey of naval architecture. Focuses
        on the 19th century. Discusses the role of sci-
        ence in the development of naval architecture.

See also: 98, 288, 296, 307, 314, 315, 327, 351, 352,
362, 373, 385, 491, 524, 535, 565, 580, 640.

AERONAUTICS AND AIRCRAFT DESIGN

Primary Sources

1443.   Beyer, Robert T., ed. **Foundations of High Speed
        Aerodynamics.** New York: Dover, 1951.

        Contains nineteen papers on high-speed aero-
        dynamics reprinted from scientific journals.
        Also contains a bibliography compiled by George
        F. Carrier.

1444.   Cayley, George. "On Aerial Navigation."
        Nicholson's Journal 24 (1809): 164-174; 25
        (1810): 81-87; 161-69.

        Triple paper written by Cayley in 1809. Laid
        the foundation for aerodynamics. Reprinted in
        the **Philosophical Magazine** 47 (1816): 81-86;
        321-29; 50 (1817): 27-35; and extended by a
        fourth paper entitled "Practical Remarks on
        Aerial Navigation." 26 (1837): 418-28. The
        triple paper is also reprinted in Gibbs-Smith
        (item 1477).

1445.   Eiffel, Gustave. **La résistance de l'air et
        l'aviation, expériences effectuées au labora-
        toire du Camp-de-Mars.** Paris: H. Dunod & E.
        Pinat, 1911.

Fundamental work on the resistance of air. Translated into English as **The Resistance of Air and Aviation. Experiments Conducted at the Champ-de-Mars Laboratory.** London: Constable & Co., 1913.

1446.   _____. **La résistance de l'air; examen des formules et des expériences.** Paris: H. Dunod & E. Pinot, 1910.

Studies on the resistance of air.

1447.   _____. **Recherches expérimentales sur la résistance de l'air exécutées à la tour Eiffel.** Paris: L. Maretheux, 1907.

Study of the resistance of air, conducted at the Eiffel Tower.

1448.   Glauert, Hermann. **The Elements of Aerofoil and Airscrew Theory.** Cambridge: Cambridge University Press, 1926.

Classic textbook on subsonic aerodynamics. Helped to make Prandtl's theory available to English-speaking aerodynamicists.

1449.   Jones, B.M. "The Streamline Aeroplane." **Journal of the Royal Aeronautical Society** 33 (1929): 357-85.

Important paper introducing engineers to the losses resulting from turbulent drag and the necessity of streamlining aircraft.

1450.   Joukowski (Zhukovskii), Nikolai E. **Bases théoriques de l'aéronautique.** Translated by S. Drzewiecki. Paris: Gauthier-Villars, 1916.

French translation of lectures by a leading theoretician of aerodynamics, given at the Imperial School of Technology in Moscow.

1451.   _____. **Geometriche Untersuchungen über die Kutta'sche Strömung.** 2 volumes. Moscow: Moscow University, 1910-11.

Helped establish the mathematical theory of circulation around an aerofoil. Resulted in the Kutta-Joukowski theorem which showed that lift was the result of lower pressure above the wing

rather than the impact of air below the wing.
(See item 1453).

1452.   _____. **Théorie tourbillonnaire de l'hélice
         propulsive.** Translated by A. Apostol.
         Paris: Gauthier-Villars, 1929.

         Study of the vortex motion of propellers.

1453.   Kutta, Wilhelm Martin. **Über ebene zirkula-
         tionsströmungen nebst flügtechnischen
         anwendungen.** Munich: Königlich Bayerischen
         Akademie der Wissenschaften, 1911.

         Contribution to the theory of circulation
         around an aerofoil. Led to the Kutta-Joukowski
         theorem concerning the calculation of the lift
         of an aerofoil. Made the study of lift a
         mathematical subject. (See item 1451).

1454.   Lanchester, Frederick W. **Aerodynamics.** London:
         Constable, 1907.

         Popular work on the theory of aerodynamics.
         Anticipated many of the ideas of Prandtl but the
         work was more qualitative than quantitative.

1455.   Lilienthal, Otto. "The Problem of Flying and
         Practical Experiments in Soaring." **Annual
         Report of the Smithsonian Institution**
         (1893): 189-99.

         Translated abstract from a paper published in
         German in **Prometheus** 4 no. 205 (1893): 769-774.
         Important contribution to the theory of
         controlled flight by one of the leaders in the
         field.

1456.   _____, with Gustav Lilienthal. **Der vogelflug
         als grundlage der fliegekunst.** Berlin: R.
         Gaertner, 1889.

         Fundamental work on the theory of controlled
         flight. Translated into English by A.W.
         Isenthal as **Birdflight as the Basis of Aviation.**
         London: Longmans, Green, & Co., 1911.

1457.   Poisson, Siméon Denis. **Recherches sur le mouve-
        ment des projectiles dans l'air, en ayant
        égard à leur figure et leur rotation, et à
        l'influence du mouvement diurne de la terre.**
        Paris: Bachelier, 1839.

        Important work on the effects of air on pro-
        jectiles by a leading French theorist.

1458.   Prandtl, Ludwig. **Abriss der strömungslehre.**
        Braunschweig: F. Vieweg, 1931.

        Fundamental work on hydrodynamics and aerody-
        namics by the teacher of Von Karman.

1459.   _____. **Application of Modern Hydrodynamics to
        Aeronautics.** Washington, D.C.: National Ad-
        visory Committee for Aeronautics, 1921.

        Essential work on the application of hydrody-
        namics to aeronautics. Helped to introduce im-
        portance of boundary layers.

1460.   _____. **Führer durch die Strömungslehre.**
        Braunschweig: F. Vieweg, 1942.

        Highly influential work on the application of
        fluid dynamics to hydraulics, aeronautics, and
        meteorology. Translated into English as **Essen-
        tials of Fluid Dynamics.** New York: Hafner, 1952.
        Also translated into French and Russian.

1461.   _____, and Oskar Karl Gustav Tietjens. **Hydro-
        und Aeromechanik nach vorlesungen von L.
        Prandtl.** Berlin: J. Springer, 1929-31.

        Work based on the lectures of Ludwig Prandtl.
        Translated into English as **Fundamentals of
        Hydro- and Aeromechanics.** New York: McGraw-Hill,
        1934.

1462.   Rateau, Auguste. **Théorie des hélices propul-
        sives marines et aériennes et des avions en
        vol rectiligne.** Paris: Gauthier-Villars,
        1920.

        Theory of propellers and theory of aerodynam-
        ics by a noted engineering scientist.

1463.   Robins, Benjamin. **New Principles of Gunnery:**
        Containing the Determination of the Force of
        Gun-Powder, and an Investigation of the Dif-
        ference in the Resisting Power of the Air to
        Swift and Slow Motion. London: J. Nourse,
        1742.

        One of the first experiments to show that
        different shaped objects with the same cross-
        sectional area had different air resistance.

1464.   Von Karman, Theodore. **Aerodynamics: Selected**
        Topics in the Light of Their Historical De-
        velopment. Ithaca: Cornell University Press,
        1954.

        A vital work by a leader in the field.

1465.   _____, and C.B. Millikan. **On the Theory of**
        Laminar Boundary Layers Involving Separation.
        Washington, D.C.: National Advisory Committee
        for Aeronautics, 1934.

        Fundamental work on laminar flow.

1466.   Wood, Ronald McKinnon. **Summary of Present State**
        of Knowledge with Regard to Airscrews.
        London: HMSO, 1919.

        Report on the theory of propellers. Helped
        to synthesize a momentum theory and a blade-ele-
        ment theory of the action of propellers. Influ-
        enced by the work of Prandtl.

See also: 395, 396, 779, 787, 1292.

Secondary Sources

1467.   Aleksandrowicz, R. **Podstawy i rozwoj lotnictwa.**
        Warsaw: Naukowo-Techniczne, 1967.

        Provides a history of aeronautics.

1468.   Becker, John V. **The High-Speed Frontier: Case**
        Histories of Four NACA Programs, 1920-1950.
        Foreword by William S. Aiken, Jr. Washington,
        D.C.: NASA Scientific and Technical Informa-
        tion Branch, 1980.

Presents histories of four aeronautical research programs including high-speed airfoils,
transonic wind tunnels, high-speed propellers,
and high-speed cowlings. Provides examples of
the creation of engineering knowledge and general design formulae. Shows that experimental
technologies, such as wind tunnels, guided the
direction of research. Concludes that progress
resulted when theory, empiricism and testing interacted.

1469.   Cohen, Robert S., and Raymond J. Seeger, eds.
        Ernst Mach: Physicist and Philosopher.
        Dordrecht: Reidel, 1970.

        Series of papers on Mach. Includes a study
        of his role in gas dynamics and shock waves.

1470.   Constant, Edward W., II. "A Model for Techno-
        logical Change Applied to the Turbojet Revo-
        lution." **Technology and Culture** 14 (1973):
        553-72.

        Earlier version of his book (item 1471).
        Concludes that technological research is work on
        a paradigm other than one's own.

1471.   _____. **The Origins of the Turbojet Revolution.**
        Baltimore: Johns Hopkins University Press,
        1980.

        Focuses on the relationship of aerodynamics
        to classical hydrodynamics. Argues for a theory
        of technological revolution based on a
        "presumptive anomaly." Believes that radically
        new technological systems replace "normal tech-
        nologies" when scientific assumptions lead to a
        "presumptive anomaly," which is the recognition
        that the system will fail at some future time.
        Significant discussion of the history and theory
        of the turbine from the 17th century to the pre-
        sent. Important work.

1472.   Dryden, Hugh L. "Fifty Years of Boundary-Layer
        Theory and Experiment." **Science** 121 (1955):
        375-80.

        Surveys developments in the history of bound-
        ary-layer theory.

1473. Durand. W.R. "Historical Sketch of the Develop-
       ment of Aerodynamical Theory." **Transactions
       of the American Association of Mechanical En-
       gineers** Aer-51-3 (1929): 13-19.

       Useful survey of the history of aerodynamics.

1474. Emme, Eugene M., and Hugh L. Dryden. **Aeronautics
       and Astronautics: An American Chronology of
       Science and Technology in the Exploration of
       Space, 1915-1960.** Foreword by Hugh L. Dryden.
       Washington, D.C.: Government Printing Office,
       1961.

       Discusses the development of aeronautics from
       the founding of NACA to the first three years of
       the space age. Includes discussion of the con-
       tributions of Von Karman, Prandtl and Goddard.

1475. Giacomelli, R., and E. Pistolesi. "Historical
       Sketch." **Aerodynamical Theory.** Edited by
       W.R. Durand. Berlin: Springer, 1934, pp.
       310-12.

       Useful survey of the history of the theory of
       aerodynamics.

1476. Gibbs-Smith, Charles H. **The Aeroplane: An His-
       torical Survey.** London: Science Museum Hand-
       book, 1960.

       Surveys the history of aircraft design. Em-
       phasizes the role of George Cayley in developing
       the theory of flight.

1477. _____. **Sir George Cayley's Aeronautics, 1796-
       1855.** London: Science Museum, 1962.

       Provides a study of an early contribution to
       the theory of flight and aircraft design. In-
       cludes reprints of some of Cayley's most impor-
       tant papers on stability of flight.

1478. Grigorian, A.T. "The Contribution of Russian
       Scientists to the Development of Aerodynam-
       ics." **Proceedings of the 10th International
       Congress of the History of Science.** Paris:
       Hermann, 1964, pp. 793-96.

1479.   Hanieski, John F.   "The Airplane as an Economic
        Variable: Aspects of Technological Change in
        Aeronautics, 1903-1955." **Technology and Cul-
        ture** 14 (1973): 535-52.

        Discusses the role of engineering theory in
        the development of fuselage design for super-
        sonic flight.

1480.   Hanle, Paul A. **Bringing Aerodynamics to Amer-
        ica.** Cambridge, Mass.: M.I.T. Press, 1982.

        Describes Theodore Von Karman's decision to
        come to Caltech from Germany. Analyzes the role
        played by Caltech president Robert Millikan and
        aircraft businessmen in attracting Von Karman to
        the Guggenheim Aeronautical Laboratory at Cal-
        tech. Provides an example of the interaction
        between science, technology and institutions.

1481.   Hooven, Frederick J.   "The Wright Brothers'
        Flight-Control System." **Scientific American**
        239 (November 1978): 167-84.

        Describes the reasons that the Wright
        Brothers' Flyers had an elevator in the front
        instead of in the rear.

1482.   Lukasiewicz, Julius.   "Canada's Encounter with
        High-Speed Aeronautics." **Technology and Cul-
        ture** 27 (1986): 223-61.

        Discusses the development of high-speed aero-
        nautics in Canada.

1483.   Owner, F.M.   "Bristol Gas Turbines--The First
        Decade." **Journal of the Royal Aeronautical
        Society** 67 (July 1963): 427-36.

        Discusses the use of gas turbines in aeronau-
        tics.

1484.   Pritchard, J.L.   "The Dawn of Aerodynamics."
        Journal of the Royal Aeronautical Society 61
        (1957): 152-56.

        Surveys the early years of aerodynamical
        theory.

1485.   _____. "Sir George Cayley (1773-1857): A Pio-
         neer in Science and Engineering." **Journal of
         the Royal Society of Arts** (February 1958):
         197-212.

         Discusses the role of Cayley in developing a
         theory of flight.

1486.   Rae, John B. "Science and Engineering in the
         History of Aviation." **Technology and Culture**
         2 (1961): 391-99.

         Discusses the role of science and engineering
         in the development of aviation.

1487.   Roland, Alex. **Model Research: The National Ad-
         visory Committee for Aeronautics, 1915-1958.**
         2 Volumes. Washington, D.C.: Scientific and
         Technical Information Branch, National Aero-
         nautics and Space Administration, 1985.

         Provides a study of the political and admin-
         istrative history of the NACA. Argues that NACA
         served the interests of the military and aero-
         nautics industry instead of conducting pure sci-
         entific research. Classifies many elements of
         engineering science as pure empiricism.

1488.   Seeger, Raymond John. "On Aerophysics Re-
         search." **American Journal of Physics** 19
         (1951): 459-69.

         Outlines the history of the effects of air on
         projectile motion.

1489.   Tani, Itiro. "History of Boundary-Layer The-
         ory." **Annual Review of Fluid Mechanics** 9
         (1972): 87-111.

         Surveys the history of the theory of boundary
         layers.

1490.   Vincenti, Walter G. "How Did It Become
         'Obvious' that an Airplane Should Be Inher-
         ently Stable?" **American Heritage of Inven-
         tion & Technology** 4 (Spring/Summer, 1988):
         50-56.

Discusses the conflicts between designing a
plane that was stable and a plane that had a
high degree of control. Concludes that piloting
tasks helped shape the debate. Argues that the
evolution of a stable, but not too stable, plane
was the result of the generation of engineering
knowledge and cannot be classified as scientific.

1491.   _____. "The Air-Propeller Tests of W.F. Durand
          and E.P. Lesley: A Case Study in Technologi-
          cal Methodology." **Technology and Culture** 20
          (1979): 712-51.

Discusses the tests of Durand and Lesley done
from 1916 to 1926. Argues that the tests represent a methodology used by engineers. Traces
the methodology back to John Smeaton in the 18th
century. Concludes that technological methods
of research differ in both form and object from
those used in the physical sciences. Important
article.

1492.   _____. "The Davis Wing and the Problem of Air-
          foil Design: Uncertainty and Growth in Engi-
          neering Knowledge." **Technology and Culture**
          27 (1986): 717-758.

Provides a case study of the relationship between engineering knowledge and design by study-
ing the adoption of the Davis wing for the B-24.
Argues that engineers must make decisions in the
face of incomplete or uncertain knowledge. Also
investigates how the demand of design led to engineering knowledge. Argues that the need to
reduce uncertainty in design is a driving force
in the growth of engineering knowledge. Important study.

1493.   Wegener, Peter P. "The Science of Flight."
          American Scientist 74 (1986): 268-78.

Surveys the 20th-century developments in the
science of aeronautics.

1494.   Wolko, Howard S., ed. **The Wright Flyer: An En-
          gineering Perspective.** Washington, D.C.:
          Smithsonian Institution Press, 1987.

Contains papers which analyze the engineering and areodynamical aspects of the Wright Flyer. Includes John D. Anderson's "The Wright Brothers: The First True Aeronautical Engineers," F.E.C. Culick and Henry R. Jex's "Aerodynamics, Stability, and Control of the 1903 Wright Flyer," Frederick J. Hooven's "Longitudinal Dynamics of the Wright Brothers' Early Flyers: A Study in Computer Simulation of Flight," Harvey H. Lippincott's "Propulsion Systems of the Wright Brothers," and Howard S. Wolko's "Structural Design of the 1903 Wright Flyer."

See also: 89, 141, 263, 338, 339, 360, 361, 383, 557, 565, 581, 604, 616.

NAME AND AUTHOR INDEX

Gibbons, J., 155
Gibbs, J. Willard, 1067,
  1099, 1100, 1222
Gibbs-Smith, Charles H.,
  603, 604, 1444, 1476,
  1477
Gibson, Arnold Hartley,
  1287, 1345
Giffard, Henri, 1223,
  1224
Gilfillan, S.C., 1428
Gille, Bertrand, 24, 605,
  606
Gille, Paul, 1429, 1430
Gillespie, William
  Mitchell, 734
Gillispie, Charles C.,
  25, 26, 281, 333
Gillispie, E.S., 334
Gillmor, C. Stewart, 335
Girard, Pierre-Simon, 331
Girault, Charles
  François, 1008
Girill, T.R., 887
Gispen, Corneliss W.R.,
  455, 541
Glauert, Hermann, 1448
Glossop, Rudolph, 815
Gnudi, M.T., 598
Goddard, Robert H., 59,
  1474
Goldbeck, Gustav, 607
Golder, H.Q., 816
Goodeve, T.M., 735, 1009
Goodman, D.C., 62
Goodstein, Judith R., 89
Goodwin, Jack, 114
Gordon, Dane R., 456
Gordon, Lewis D.B., 736,
  737, 791
Gordon, Robert B., 1346
Grashof, Franz, 359, 825,
  852, 1010
Grattan-Guinness, I.,
  336, 457
Gray, John Macfarlane,
  1101, 1394
Grayson, Lawrence P., 458
Greene, Charles E., 913
Greene, John C., 156

Gregory, Olinthus
  Gilbert, 1011
Grigorian, A.T., 1478
Groner, Alex, 527
Gross, Walter E., 1059
Grosser, Morton, 337
Grove, J.W., 157
Gruber, W.H., 236, 243
Gruender, C.D., 158
Guényveau, André, 1288
Guericke, Otto von, 608,
  676
Guillerme, André, 1347
Gutting, Gary, 159
Guyonneau de Pambour,
  François Marie, 1102,
  1103, 1104

Hâchette, Jean Nicolas
  Pierre, 928, 1012, 1105
Hacker, Barton C., 59,
  609
Hahn, Roger, 542
Hall, A. Rupert, 66, 160,
  161, 162, 163, 164,
  459, 543, 610, 611,
  612, 1431
Hall, Bert S., 613
Hall, Marie Boas, 614,
  615
Hall, Michael, 7
Hall, R. Cargill, 338
Halle, Gerhard, 339
Hamilton, Stanley B.,
  817, 966
Hanieski, John F., 1479
Hankins, Thomas L., 340
Hanle, Paul A., 1480
Harrigan, Patrick J., 423
Harris, F.R., 1207
Harrison, James, 341
Hart, Clive, 616
Hart, Ivor B., 617, 618
Hartenberg, Richard S.,
  1060, 1061
Haton de la Goupillière,
  Julien Napoleon,
  1013,1014
Haupt, Herman, 342, 384,
  914, 915
Hays, J.N., 343

Reynolds, Osborne, 289,
 334, 403, 779, 1036,
 1052, 1137, 1138, 1281,
 1293, 1345, 1367, 1440
Reynolds, Terry S., 1357,
 1358
Rezneck, Samuel, 489
Rich, Elihu, 1079
Richardson, George, 982
Ricketts, Palmer C., 490
Rink, Evald, 103, 671
Ripper, William, 1039,
 1139, 1140
Ritchie-Calder, Peter,
 566
Robertson, Paul L., 491,
 1436
Robins, Benjamin, 1463
Robinson, Eric, 229, 367,
 404, 1248
Robison, John, 780, 866,
 935
Roderick, Gordon W., 492,
 493, 494, 495, 496
Rodman, Thomas Jackson,
 877, 878
Rogers, William Barton,
 867
Roland, Alex, 1487
Rolt, L.T.C., 62, 310,
 368, 374, 567
Root, J.W., 957
Rose, Paul Lawrence, 672,
 673
Rosenberg, Nathan, 244,
 990
Rosman, Holgar, 369
Rossi, Paolo, 674
Rothenberg, Marc, 104,
 199, 398, 402
Rouse, Hunter, 105, 1359,
 1360
Rousseau, G.S., 132, 1179
Roysdon, Christine M.,
 114, 286
Ruddock, Ted, 983
Rumford, Benjamin
 Thompson, Count, 308,
 375, 405, 545
Rumpf, H., 245
Rürup, Reinhard, 497

Russell, Colin A., 63
Russell, John Scott, 327,
 1141, 1405, 1412, 1413,
 1414, 1415, 1416, 1417
Russo, François, 64, 108

Saint-Venant, Adhémar
 Jean Claude Barré de,
 781, 782, 833, 844,
 868, 869, 870, 871,
 1319
Salvadori, Mario, 984
Sanderson, Michael, 498
Sandler, Stanley, 1438
Sarton, George, 109
Sass, Friedrich, 1249
Sattiram, Kappal, 1439
Saunders, O.A., 246
Sawers, David, 32
Scaglia, Gustina, 656,
 657, 675
Scaife, W. Garrett, 1250
Schimank, Hans, 676
Schneer, Cecil J., 823
Schneider, Ivo, 65
Schnitter, N.J., 110
Schofield, Robert E.,
 247, 568, 569, 570
Schriff, Edmund, 813
Schwamb, Peter, 1040
Scott, John C., 1440
Scott, Quinta, 977
Sebestik, Jan, 248
Seeger, Raymond John,
 1469, 1488
Seely, Bruce E., 571
Semenev, G.I., 249
Sennett, Richard, 1142
Shapin, Steven, 499
Sharlin, Harold Issadore,
 370
Sharlin, Tiby, 370
Shaw, John, 1361
Shelby, Lon R., 677
Sheriff, Thomas, 1251
Sherwin, C.W., 250
Shinn, Terry, 500, 501,
 572
Shipley, William, 287
Show, Ralph R., 111
Shukhardin, S.V., 219